Collins

AQA GCSE 9-1
Maths
Foundation

Workbook

Phil Duxbury and Trevor Senior

Preparing for the GCSE Exam

Revision That Really Works

Experts have found that there are two techniques that help you to retain and recall information and consistently produce better results in exams compared to other revision techniques.

It really isn't rocket science either – you simply need to:

- **test yourself** on each topic as many times as possible
- **leave a gap** between the test sessions.

Three Essential Revision Tips

1. Use Your Time Wisely

- Allow yourself plenty of time.
- Try to start revising at least six months before your exams – it's more effective and less stressful.
- Don't waste time re-reading the same information over and over again – it's not effective!

2. Make a Plan

- Identify all the topics you need to revise.
- Plan at least five sessions for each topic.
- One hour should be ample time to test yourself on the key ideas for a topic.
- Spread out the practice sessions for each topic – the optimum time to leave between each session is about one month but, if this isn't possible, just make the gaps as big as you can.

3. Test Yourself

- Methods for testing yourself include: quizzes, practice questions, flashcards, past papers, explaining a topic to someone else, etc.
- Don't worry if you get an answer wrong – provided you check what the correct answer is, you are more likely to get the same or similar questions right in future!

Visit **collins.co.uk/collinsGCSErevision** for more information about the benefits of these techniques, and for further guidance on how to plan ahead and make them work for you.

Command Words used in Exam Questions

This table explains some of the most commonly used command words in GCSE exam questions.

Command word	Meaning
Calculate...	A calculator or a formal method is needed
Estimate...	Round values to 1 s.f. in obtaining your answer
Explain...	Use words to explain your answer
Factorise...	Take out the common factor or factorise into two brackets if a quadratic
Factorise fully...	Indicates that there are at least two stages in the factorisation
Prove...	An algebraic, or geometric, proof is required
Show that...	Use words, numbers or algebra to show the answer
Simplify...	Collect terms together
Simplify fully...	Collect terms together and factorise the answer
Solve...	Find the value(s) of x that makes the equation true
Work out...	A calculation is involved but it may be possible to do it mentally
Write down...	Working is not needed to find the answer

Contents

 N Number   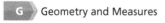 **A** Algebra   **G** Geometry and Measures

 S Statistics   **P** Probability   **R** Ratio, Proportion and Rates of Change

Number 1, 2 & 3

Grade 1-3 **1** Work out $3 + 12 - 2 \times 3$

_____ [1]

Grade 1-3 **2** Calculate $\dfrac{1}{6} + \dfrac{2}{3}$

_____ [1]

Grade 1-3 **3** Work out 0.015×0.3

_____ [1]

Grade 1-3 **4** Calculate 79×23

_____ [1]

Grade 1-3 **5** Consider this list of numbers. 10, 14, 24, 27, 30, 36, 37, 48

From the above list, write down the number that is:

a) a prime number _____ [1]

b) a square number _____ [1]

c) a multiple of both 3 and 5 _____ [1]

d) the lowest common multiple of 6 and 8 _____ [1]

e) a factor of 42 _____ [1]

Grade 1-3 **6** Work out

a) $(-12) + (-3)$ _____ [1]

b) $3 - (-12)$ _____ [1]

c) $(-3) \times (-12)$ _____ [1]

Grade 1-3 **7** Six family members went on a trip to the cinema. Each ticket cost £6.50 per person.

Calculate the total change received from a £50 note.

£ _____ [2]

8 The blades of a wind turbine turn at a rate of 18 revolutions per minute.

How many complete rotations will the blades make in 3 minutes and 20 seconds?

_____ [2]

9 Work out $(3.6 \times 10^3) - (1.1 \times 10^2)$

Give your answer in standard form.

_____ [2]

10 The distance between the Earth and the Sun is 147 million kilometres.

Convert this distance to metres. Give your answer in standard form.

_____ m   [2]

11 It is given that $5 + \dfrac{24}{x}$ is an integer less than 8.

Work out two possible integer values for x.

$x =$ _____ or $x =$ _____   [2]

12 Write 84 as the product of prime factors.

_____ [2]

13 Work out $(5 \times 10^5)^2$
Give your answer in standard form.

_____ [2]

14 Work out $(-2.5)^3$
Give your answer as a fraction.

_____ [2]

Total Marks _____ / 28

Basic Algebra

Grade 1–3   **1**   Solve the equation $\frac{x}{9} = 3$

$x =$ _____ [1]

Grade 1–3   **2**   Solve the equation $8x - 5 = 27$

$x =$ _____ [2]

Grade 1–3   **3**   Simplify $y - 8x + 2y$

_____ [1]

Grade 1–3   **4**   Simplify $12p + 4q - 3p - 7q$

_____ [1]

Grade 3–5   **5**   Expand the brackets and simplify $7(2y - 3) + 4(3y + 2)$

_____ [2]

Grade 3–5   **6**   You are given that $a = 2$ and $b = -5$

Work out the value of $a^2 + b^2$

_____ [2]

Grade 3–5   **7**   Solve the equation $6 - 2(2 - x) = 20$

$x =$ _____ [2]

Grade 3–5   **8**   Solve the equation $3(x - 2) = 12(7 - x)$

$x =$ _____ [3]

Total Marks _____ / 14

Factorisation and Formulae

Grade 3–5 **1** Factorise completely $27p^3q - 18q^2$

_____ [1]

Grade 3–5 **2** Expand $(x + 10)(x + 2)$ 🖩

_____ [2]

Grade 3–5 **3** Factorise $x^2 - 36$ 🖩

_____ [1]

Grade 3–5 **4** A party is held in a large marquee. The cost (£C) of renting the marquee over a period of time is given by $C = 150 + 12t$, where t is the time in hours.

a) Work out the cost of renting the marquee for a $5\frac{1}{2}$ hour period.

£ _____ [1]

b) Rearrange the formula to make t the subject.

_____ [2]

c) A customer has £350. Work out the longest time, to the nearest hour, for which he can rent the marquee.

_____ [3]

Grade 3–5 **5** Factorise $x^2 + x - 42$

_____ [2]

Grade 3–5 **6** Rearrange $5(x - 2) = 2q + 10$ to make x the subject.

_____ [3]

Total Marks _____ / 15

Topic-Based Questions

Ratio and Proportion

1 Express 10 : 5 in the ratio 1 : n

Circle your answer.   2 : 1   1 : 2   2 : $\frac{1}{2}$ 1 : $\frac{1}{2}$ [1]

2 Shop A sells 12 bars of Chocomunch for £7.50
Shop B sells 9 bars of Chocomunch for £5.75

Which shop offers the better value?

_____ [2]

3 Divide £360 in the ratio 3 : 5

_____ [3]

4 Express the ratio 27 : 18 : 90 in its simplest form.

_____ [1]

5 Two quantities x and y are related by the formula $y = \frac{4}{5}x$

Work out the ratio $y : x$, expressing your answer in the form 1 : n

_____ [1]

6 To make a light shade of orange paint, red paint and yellow paint are mixed in the ratio 5 : 8

If Helen has 12 litres of red paint, how much yellow paint will she need?

_____ l [2]

7 A map of Cyprus is produced using a scale of 1 : 250 000
The distance in real life between Paphos and Limassol is 57 km.

How far apart (in centimetres) would these towns appear on the map?

_____ cm [3]

Total Marks _____ / 13

Variation and Compound Measures

Grade 3–5 **1** y is directly proportional to x.

What happens to y when x is doubled? Circle your answer.

y stays the same y is halved y is doubled y is squared   [1]

Grade 3–5 **2** The speed limit on a dual carriageway is 70 mph. A speed of 1 m/s is equivalent to 2.237 mph.

Convert 70 mph to m/s. Give your answer to 1 decimal place.

_____ m/s   [2]

Grade 3–5 **3** A grizzly bear weighs 2500 N. Work out the pressure the bear exerts on the floor of a rectangular cage of area 45 m².

_____ N/m²   [1]

Grade 3–5 **4** The density of zinc is 7.13 g/cm³. Work out the volume present in a 0.15 kg bar of zinc.

_____ cm³   [3]

Grade 3–5 **5** $y = kx$ When $x = 8$, $y = 16$

a) Work out the value of k.

$k = $ _____   [1]

b) Work out the value of x when $y = 25$.

$x = $ _____   [2]

Grade 3–5 **6** $y = \dfrac{k}{x}$ When $x = 4$, $y = 5$

a) Work out the value of k.

$k = $ _____   [1]

b) Work out the value of y when $x = \dfrac{1}{2}$

$y = $ _____   [2]

Total Marks _____ / 13

Angles and Shapes 1 & 2

1 Work out the value of x in the following diagram.

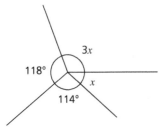

$x =$ _____ degrees   [3]

2 The following diagram *ABCDE* shows a regular pentagon.

a) Work out the sum of the interior angles of the pentagon.

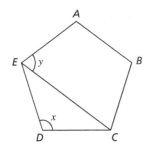

_____ degrees   [1]

b) Without using a protractor, work out the size of angle x and the size of angle y.

$x =$ _____ degrees

$y =$ _____ degrees   [3]

3 Work out the size of the angle marked x.

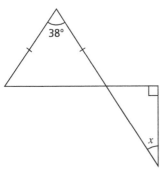

$x =$ _____ degrees   [3]

4 Work out the sum of the interior angles of a decagon.

_____ degrees   [2]

Total Marks _____ / 12

Fractions

Grade 1-3

1 Work out $\frac{2}{5}$ of £70

£ _____ [1]

Grade 1-3

2 Write the following fractions in ascending order: $\frac{3}{4}, \frac{21}{32}, \frac{5}{8}, \frac{13}{16}, \frac{1}{2}$

_____ [2]

Grade 1-3

3 Eddie's English homework was to finish writing a 1500-word essay on a Shakespearean play. He had already written 350 words in class.

What fraction of the essay did he have left to complete?
Express your answer in its simplest form.

_____ [2]

Grade 3-5

4 Lexie ate $\frac{1}{6}$ of a large box of chocolates. She then noticed there were only 60 chocolates left.

How many chocolates were there in the box originally?

_____ [2]

Grade 3-5

5 Work out $\frac{1}{2}$ of $\frac{3}{4}$ of 64

_____ [2]

Grade 3-5

6 Work out $3\frac{1}{4} \div 1\frac{2}{5}$

Give your answer as a mixed number.

_____ [4]

Total Marks _____ / 13

Percentages 1 & 2

Grade 1–3

1 Work out the percentage increase from 32 to 75. Give your answer to 1 decimal place.

_____ %   [2]

Grade 1–3

2 Work out 900 ml as a percentage of 1.5 *l*.

_____ %   [3]

Grade 3–5

3 Bob the taxi driver's annual gross salary is £24 000.
He pays tax at a rate of 18% on all of his income.

a) Calculate Bob's monthly pay before tax.

£ _____   [1]

b) Calculate Bob's monthly pay after tax.

£ _____   [2]

Grade 3–5

4 Ronnie bought an antique statue for £1400. Some years later, Ronnie had the statue valued and was informed that its value had increased by 125%.

Work out the current value of the statue.

£ _____   [3]

Grade 3–5

5 There were an estimated 1995 wild lemurs in Madagascar in the year 2020, a decrease of 95% compared to the year 2000.

How many lemurs were present in the year 2000?

_____   [2]

Total Marks _____ / 13

Probability 1 & 2

1 For each of the following events, circle the correct probability from the given options.

a) A fair dice is rolled and the outcome is an even number.

$\frac{1}{4}$ $\frac{1}{2}$ $\frac{3}{4}$ 1   [1]

b) Two fair coins are tossed at the same time and the outcome is two heads.

10%   25%   40%   50%   [1]

c) A five-sided spinner with equally-sized segments numbered 1 to 5 is spun and the outcome is a prime number.

0.2   0.4   0.5   0.6   [1]

2 A bag contains 20 coloured counters: 7 red, 9 green and 4 yellow.
A counter is selected at random.

a) Work out the probability that it is either red or yellow.

_____ [1]

b) Work out the probability that the counter is not red.

_____ [1]

c) State the probability that the counter is orange.

_____ [1]

3 A spinner has four numbers: 2, 3, 4 and 6. Margie spins the spinner twice and adds the numbers together to obtain a final score.

a) Complete the following table showing the possible scores.   [2]

2nd spin

		2	3	4	6
	2	4			
1st spin	3				
	4				
	6				

b) Work out the probability that Margie scores 9.

_____ [1]

c) Work out the probability that Margie gets a score greater than 5.

_____ [1]

4 Dale takes a Maths exam and an English exam.

The probability that he passes Maths is 0.8
The probability that he passes English is 0.6

a) State the probability that he does not pass his Maths exam.

_____ [1]

b) Complete the tree diagram by adding the correct probabilities to each branch.

[2]

c) Work out the probability that Dale passes only one of these exams.

_____ [3]

5 Jemma has a two-sided coin, where the probability of throwing a head is 0.58

a) Explain why the coin is biased.

_____

_____ [1]

b) If Jemma throws the coin 250 times, how many times would she expect it to land on tails?

_____ [2]

Total Marks _____ / 19

Number Patterns and Sequences 1

Grade 1-3

1 The following set of numbers forms part of the sequence generated by using the term-to-term rule 'subtract 5 and multiply by 3'.

Work out the two missing numbers in the sequence below.

__, 15, 30, 75, __   _____ [2]

Grade 1-3

2 Write down the next two terms in the following sequences.

a) 1, 5 ,9, 13, __, __   _____ [2]

b) 1, 3, 6, 10, __, __   _____ [2]

c) 1, 8, 27, __, __   _____ [2]

3 An arithmetic sequence is given by 11, 14, 17, 20, …

Grade 1-3

a) State the next two terms in this sequence.

_____ [1]

Grade 3-5

b) Work out an expression for the nth term in this sequence.

_____ [2]

Grade 3-5

c) Which of these numbers would occur in this sequence? Circle the correct answer.

501   502   503   [1]

Grade 3-5

4 Write down the next two terms in the following sequences.

a) The Fibonacci sequence   7, 5, 12, 17, __, __

_____ [2]

b) The geometric sequence $\frac{2}{3}, \frac{4}{9}$, __, __

_____ [2]

Grade 3-5

5 Work out an expression for the nth term of the arithmetic sequence starting 9, 15, 21, …

_____ [2]

Total Marks _____ / 18

Number Patterns and Sequences 2

1 Here is a sequence of patterns made using sticks.

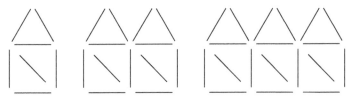

Pattern 1   Pattern 2   Pattern 3

Grade 1–3

a) Draw the next pattern in the sequence.

[1]

Grade 1–3

b) Complete the table below to show the number of sticks needed for each pattern.

Pattern number	1	2	3	4
Number of sticks	7	13		

[2]

Grade 3–5

c) Work out the position-to-term rule for this sequence.

_____ [2]

Grade 3–5

d) Is it possible to make a pattern from this sequence using exactly 150 sticks? Explain your answer.

_____

_____ [3]

Grade 3–5

2 The nth term in a sequence is given by the formula $4 - 3n$.

a) Work out the first four terms in the sequence.

_____ [2]

b) Work out the 100th term in the sequence.

_____ [1]

Total Marks _____ / 11

Transformations

1 **a)** On the grid, enlarge the given triangle by a scale factor of $\frac{1}{3}$, using the centre of enlargement (6, 1). [2]

b) Write down the coordinates of the vertices of the new triangle you have drawn.

_____ [1]

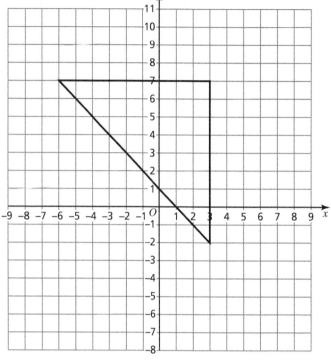

2 The grid shows three triangles.

a) Describe the transformation that maps Triangle *ABC* to Triangle *DEF*.

_____

_____

_____ [3]

b) Describe the transformation that maps Triangle *ABC* to Triangle *GHJ*.

_____

_____

_____ [2]

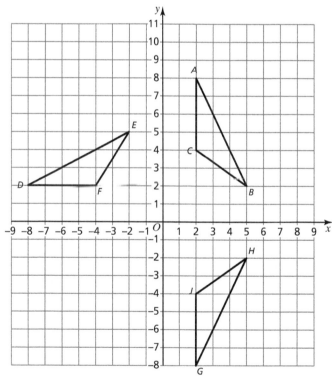

c) Ahmed says only triangles *ABC* and *DEF* are congruent, but Sunil disagrees and thinks all three triangles are congruent. Who is correct, and why?

_____

_____ [1]

Total Marks _____ / 9

Constructions

 1 The following straight line forms the base of an equilateral triangle.

Using a ruler and a pair of compasses, construct two more lines in order to complete the triangle. Show your construction lines clearly.

[3]

_____

 2 The following diagram shows a point *X* and a line *AB*.

Using a ruler and a pair of compasses, construct a perpendicular line to *AB* from the point *X*.

[3]

X
•

A ——————————————————————— B

3 The following diagram shows three points *X*, *Y* and *Z*.

Using a pair of compasses, draw the locus of points that are less than 2 cm from each point.

[3]

X
•

Y
•

•
Z

Total Marks _____ / 9

Nets, Plans and Elevations

Grade 3–5

1 **a)** Sketch a plan view of the 3D shape shown below.

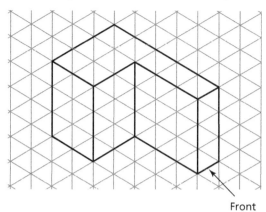

Front

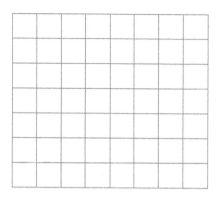

[1]

b) Sketch the front elevation of the shape.

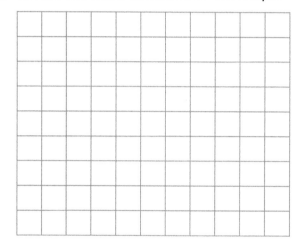

[1]

c) How many unit cubes would be required to build the above shape?

_____ [1]

Grade 3–5

2 On the isometric paper below, draw a cuboid of dimensions 8 × 3 × 2 units.

[3]

Total Marks _____ / 6

Linear Graphs

Grade 3–5 **1** The diagram shows the graph of $y = mx + 3$.

a) Work out the value of m.

_____ [2]

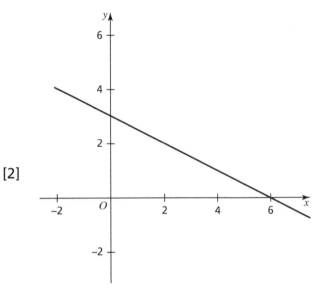

b) Does the point (–12, 10) lie above or below this line?

Give a reason for your answer.

_____

_____ [2]

Grade 3–5 **2** Which **two** of the following four lines are parallel to each other?

A $4y = 4x - 1$ **B** $y = 4x - 1$ **C** $y = \frac{1}{4}x + 1$ **D** $8x - 2y = 8$

_____ [1]

Grade 3–5 **3** A line has equation $5y - x = 6$

Work out the coordinates of the point where the graph crosses

a) the x-axis

_____ [1]

b) the y-axis

_____ [1]

Grade 3–5 **4** A line goes through the points (4, 2) and (10, 14).

Work out the equation of the line, writing your answer in the form $y = mx + c$

_____ [4]

GCSE Maths Workbook

Total Marks _____ / 11

Graphs of Quadratic Functions

Grade 3–5 **1** The diagram shows the graph of
$y = 5 - 4x - x^2$

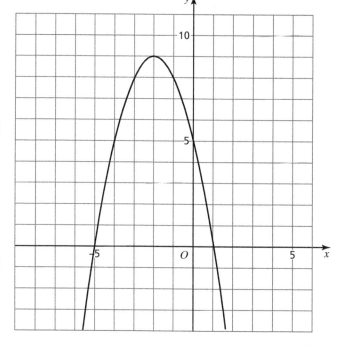

a) State the coordinates of the turning point of the graph.

_____ [1]

b) Use the graph to solve the equation
$5 - 4x - x^2 = 0$

$x =$ _____, $x =$ _____ [2]

Grade 3–5 **2** On the grid, sketch the graph of $y = x^2 - 4$ from
$x = -2$ to $x = 2$.
Show clearly where the graph crosses the axes. [3]

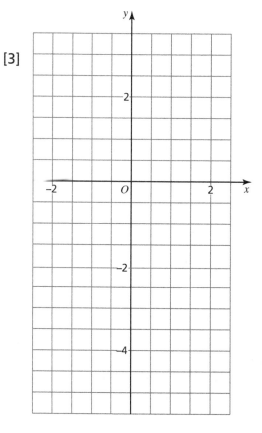

Total Marks _____ / 6

Powers, Roots and Indices

Grade 3–5 **1** a) Simplify $2p^3 \times p^4$

_____ [1]

  b) Simplify $q^8 \div q^2$

_____ [1]

  c) Simplify $(2r^3)^2$

_____ [1]

Grade 3–5 **2** Simplify $4x^3y^2 \times 5x^2y$

_____ [2]

Grade 3–5 **3** Write down these values in order of size, starting with the smallest.

  2^3 3^2 3^0 2^{-1}

_____ [2]

Grade 3–5 **4** Work out $\sqrt{81} - \sqrt{16}$

_____ [2]

Grade 3–5 **5** Write 3^{-2} as a fraction.

_____ [2]

Grade 3–5 **6** Work out $\sqrt{12^2 + 5^2}$

_____ [2]

Total Marks _____ / 13

Area and Volume 1, 2 & 3

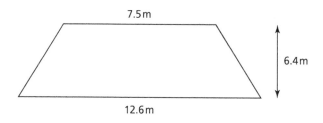

1 The diagram shows a trapezium-shaped wall.

Work out the area of the wall.
Give your answer to 3 significant figures.

7.5 m

6.4 m

12.6 m

_____ m²   [3]

2 The following diagram shows a sector of radius 20 cm.

Work out the perimeter of the sector.
Give your answer to 1 decimal place.

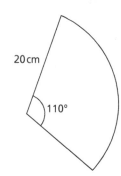

20 cm

110°

_____ cm   [4]

3 The area of a circle is 50 cm².

Work out the length of its circumference.
Give your answer to 3 significant figures.

_____ cm   [4]

4 The diagram shows six circles, each of radius 2 cm, surrounded by a rectangle.

Work out the shaded area.
Give your answer to 3 significant figures.

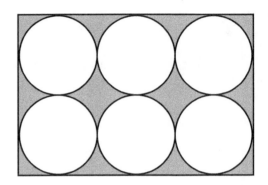

_____ cm² [4]

5 The frustum has height 3 cm and base radius 2 cm.

Work out the volume of the frustum.
Give your answer in terms of π.

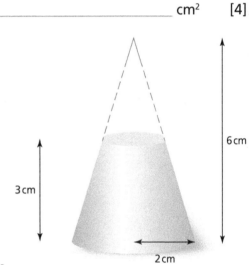

_____ cm³ [6]

6 The diagram shows a hemisphere of diameter 10 cm.

Work out the **total** surface area of the hemisphere.

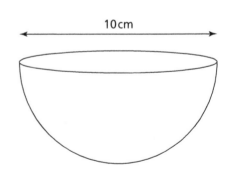

_____ cm² [3]

Total Marks: _____ / 24

Uses of Graphs

1 This diagram models how the value (£V) of a new £20 000 car varies with time (t years). The equation of the line is given by

$V = 20\,000 - 1600t$

Grade 1–3

a) Calculate the value of the car after 3 years.

£ _____ [1]

Grade 3–5

b) Write down the gradient of the line.

_____ [1]

Grade 3–5

c) What does the gradient represent?

_____ [1]

Grade 3–5

d) After how many years does the car have no value?

_____ [2]

Grade 3–5

e) Give **one** reason why the model may not be accurate in real life.

_____ [1]

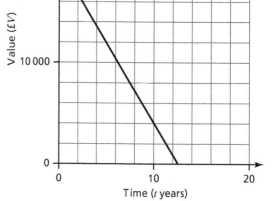

Grade 3–5

2 Which graph below shows that y is inversely proportional to x? _____ [1]

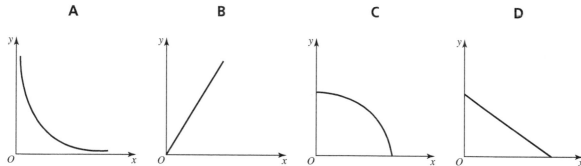

Grade 3–5

3 Mary says, "The circumference of a circle is directly proportional to its radius."

Ashraf says, "The circumference of a circle is directly proportional to its diameter."

Who is correct?

A Neither are correct   B Only Mary is correct

C Only Ashraf is correct   D Both Mary and Ashraf are correct. _____ [1]

Total Marks: _____ / 8

Other Graphs

1 The graph shows the conversion rate between British Pounds (GBP) and Polish zloty (PLN) on a certain date.

a) Use the graph to work out how many zloty could be exchanged for £12.

_____ zloty   [1]

b) Use the graph to work out how many pounds could be exchanged for 500 Polish zloty.

£ _____   [2]

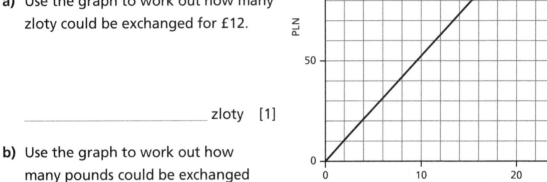

2 The diagram shows the speed–time graph of Marv the snail, travelling in a straight line, over a three-hour period.

a) Work out the acceleration of Marv over the first 90 minutes.

_____ cm/h²   [2]

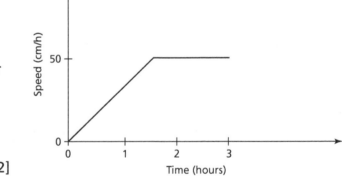

b) Describe the motion of Marv after the first 90 minutes.

_____   [1]

3 On the axes given, sketch the graph of $y = x^3$

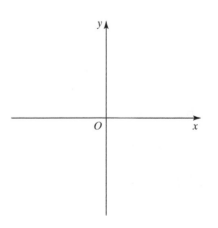

[2]

Total Marks: _____ / 8

Inequalities

1 Solve the following inequalities.

a) $x - 3 < 15$

_____ [1]

b) $3x + 2 \geqslant 14$

_____ [1]

c) $4x - 5 \leqslant 16 + x$

_____ [1]

2 $-2 < x \leqslant 3$ and x is an integer. Write down all possible values for x.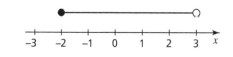

_____ [1]

3 Solve $5(x - 2) > 20$

_____ [2]

4 $-16 \leqslant 5x < 20$ and x is an integer.

Work out all possible values for x.

_____ [2]

5 Which inequality is shown on this number line?

Circle your answer.

$-2 < x < 3$ $-2 \leqslant x < 3$ $-2 < x \leqslant 3$ $-2 < x \geqslant 3$ [1]

6 Show the inequality $x > -1$ on the number line below.   [2]

7 $x > -1$ and $2x \leqslant 6$

Work out the possible integer values of x.

$x =$ _____ [3]

Total Marks: _____ / 14

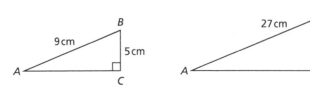

1 *ABC* and *ADE* are similar triangles.

Work out the length x.

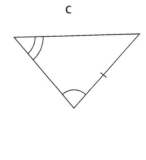

_____ cm [2]

2 Which **two** of triangles A, B and C are congruent? Explain why.

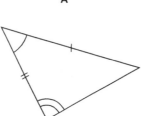

_____

_____ [2]

3 The two trapezia *ABCD* and *EFGH* are mathematically similar.

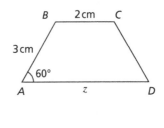 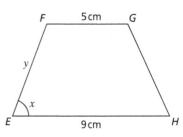

a) Write down the size of angle x.

_____ degrees [1]

b) Work out the length y.

_____ cm [2]

c) Work out the length z.

_____ cm [2]

Total Marks: _____ / 9

Right-Angled Triangles 1

1 *PQR* is a right-angled triangle.

Work out the length of *PR*. Give your answer to 3 significant figures.

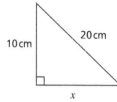

_____ cm [2]

2 Calculate the value of x in this diagram.

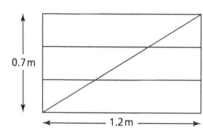

_____ cm [3]

3 The diagram shows a rectangular gate. It has seven straight pieces.

Work out the total length of the pieces.

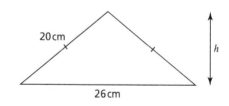

_____ m [4]

4 Triangles A, B, C and D have sides of lengths shown below.

Which of these triangles is **not** a right-angled triangle? Show your working.

A 5 cm, 12 cm, 13 cm   **B** 17 cm, 1.44 m, 1.45 m   **C** 20 cm, 23 cm, 31 cm   **D** 20 cm, 99 cm, 1.01 m

_____

_____

_____ [2]

5 The diagram shows an isosceles triangle, of perpendicular height h.

a) Calculate h. Give your answer to 3 significant figures.

$h = $ _____ cm [2]

b) Use your answer to part a) to work out the area of the triangle.

_____ cm² [2]

Total Marks: _____ / 15

Right-Angled Triangles 2

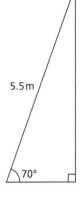

Grade 3-5

1 A ladder of length 5.5 m is placed against a vertical wall, on horizontal ground. The angle the ladder makes with the ground is 70 degrees.

Work out the distance from the foot of the ladder to the wall.

5.5 m

70°

_____ m   [2]

Grade 3-5

2 Work out the length marked x in this right-angled triangle.

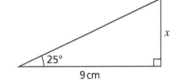

x

25°

9 cm

_____ cm   [2]

Grade 3-5

3 a) Work out the value of angle x in the triangle shown right.

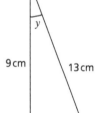

y

9 cm

13 cm

x

$x =$ _____ degrees   [2]

b) Work out the value of angle y.

$y =$ _____ degrees   [1]

Grade 3-5

4 Here are four statements.

$\sin 0° = \cos 0°$ $\sin 30° = \cos 60°$ $\sin 45° = \cos 45°$ $\sin 60° = \cos 30°$

How many of these statements are true? Circle your answer.

None of them are true   Only one of them is true

Exactly two of them are true   Exactly three of them are true   [1]

Total Marks: _____ / 8

Statistics 1

Grade 1-3

1 The pictogram shows the number of days students were absent from Trumpton High School over a particular term.

In Year 10, there were 28 days of absence recorded.

In Year 9, there were 14 days of absence recorded.

a) Work out a key for the pictogram.

[3]

Year group	
7	○ ○ ○ ○ ○
8	○ ○ ○
9	
10	○ ○ ○ ◖

b) How many days absences were recorded for Year 7 and Year 8 in total?

_____ [1]

c) Complete the pictogram for Year 9.

[2]

2 The pie chart shows the results of a music survey, where a number of students were asked to state their favourite type of music.

Fifteen students in total chose 'Pop' as their favourite type of music.

Grade 1-3

a) State the modal type of music.

_____ [1]

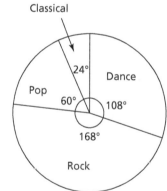

Grade 1-3

b) How many students were surveyed in total?

_____ [2]

Grade 1-3

c) How many students chose 'Dance' as their favourite music type?

_____ [2]

Grade 3-5

d) What percentage of the students surveyed preferred classical?

_____% [2]

Total Marks: _____ / 13

Statistics 2

Grade 3–5 **1** The table shows the mass of 25 cats examined by a vet over the course of one week:

Mass of cat, M kg	Number of cats	Midpoint	
$3 \leqslant M < 4$	4		
$4 \leqslant M < 5$	5		
$5 \leqslant M < 6$	9		
$6 \leqslant M < 7$	7		

Complete the table and calculate an estimate for the mean mass from this information.

_____ kg   [3]

Grade 3–5 **2** In a family of four people are mum, dad and their two children.
The two children are aged 14 and 16. Their dad is aged 41.
The mean age of the family is 27.

Calculate the age of mum.

_____   [2]

Grade 3–5 **3** The scatter graph shows the monthly rents
of various flats and their distances
from the centre of London.

a) Describe the relationship between
monthly rent and the distance
from the centre of London.

_____   [1]

b) Use the graph to estimate the cost of
renting a flat 38 km from the centre
of London.

£ _____   [2]

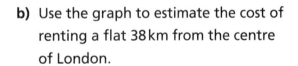

c) Give a reason why your answer to part b)
may not be reliable.

_____   [1]

Total Marks: _____ / 9

Measures, Accuracy and Finance

Grade 1–3

1 The mathematical constant π is a non-recurring decimal, and is equal to 3.141 592 653 589 793...

Write down the value of π to:

a) 3 significant figures _____ [1]

b) 3 decimal places _____ [1]

c) 5 significant figures _____ [1]

d) 5 decimal places _____ [1]

Grade 3–5

2 The average length of a car sold in the UK is 15.5 feet.

Work out the average car length in metres. Give your answer to 3 significant figures.
Assume that: 1 inch = 2.54 cm   1 foot = 12 inches

_____ m   [2]

Grade 3–5

3 Estimate the value of $\dfrac{9.09 \times 5.9}{(31 - 12.9) \times 3.1}$

_____ [2]

Grade 3–5

4 The time in Hong Kong is 7 hours ahead of the UK. The time in Chicago is 6 hours behind the UK. A businesswoman in Chicago needs to phone her colleague in Hong Kong at 9.30 am Hong Kong time.

At what time in Chicago should she pick up the phone?

_____ [2]

Grade 3–5

5 Max sold a rare CD on an internet auction site, where the winning bid was £15.26.

Given that Max made a 40% profit, how much did he originally pay for the CD?

£ _____ [2]

Total Marks: _____ / 12

Quadratic and Simultaneous Equations

1 Solve the simultaneous equations $2x + y = 5$

$8x - y = 5$

$x =$ _____

$y =$ _____ [3]

2 This is the graph of $y = 6 - x - x^2$

a) Use the graph to write down the solutions of the equation $6 - x - x^2 = 0$

$x =$ _____, $x =$ _____ [2]

b) Use the graph to solve the equation $6 - x - x^2 = 4$

$x =$ _____, $x =$ _____ [2]

3 a) Factorise $x^2 + 5x - 14$

_____ [2]

b) Solve the equation $x^2 + 5x - 14 = 0$

$x =$ _____ or $x =$ _____ [1]

Total Marks: _____ / 10

Circles

Grade 1–3

1 Fill in the gaps from the selection of words below.

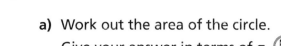

chord	radius	tangent	diameter	arc	sector

a) A chord going through the centre of a circle is called the _____. [1]

b) A line that touches the circumference of a circle only once is called a _____. [1]

2 Here is a circle of radius 4 cm.

Grade 1–3

a) Work out the area of the circle.
Give your answer in terms of π.

_____ cm² [1]

Grade 3–5

b) On the circle, draw a sector of area
2π cm². [3]

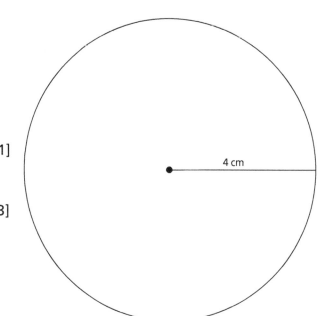

4 cm

Grade 3–5

3 In the diagram, *OAB* is a quarter-circle of radius 15 cm.

Work out the shaded area.
Give your answer to 1 decimal place.

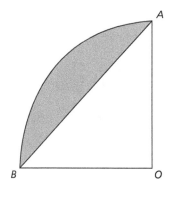

_____ cm² [6]

Total Marks: _____ / 12

Vectors

Grade 3–5 **1** The diagram shows two triangles, *ABC* and *A'B'C'*.

a) Describe the transformation of *ABC* to *A'B'C'* using a column vector.

_____

_____ [2]

b) Describe the transformation of *A'B'C'* to *ABC* using a column vector.

_____

_____ [2]

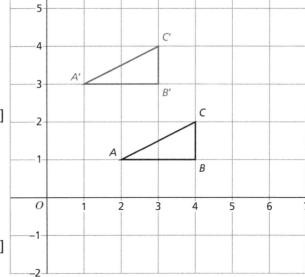

Grade 3–5 **2** $\mathbf{a} = \begin{pmatrix} 2 \\ -4 \end{pmatrix}$ $\qquad \mathbf{b} = \begin{pmatrix} -6 \\ 1 \end{pmatrix}$ $\qquad \mathbf{c} = \begin{pmatrix} 1 \\ -2 \end{pmatrix}$

a) Which two of the vectors **a**, **b** and **c** are parallel? Give a reason for your answer.

_____ [1]

b) Work out **b** + 2**c** as a column vector.

$\begin{pmatrix} \\ \end{pmatrix}$ [2]

c) Work out the value of k if $5\mathbf{a} + 2\mathbf{b} = k\begin{pmatrix} 1 \\ 9 \end{pmatrix}$

$k = $ _____ [3]

Grade 3–5 **3** The diagram shows a vector **p** and a vector **q**.

On the same diagram, draw a line with an arrow to show the vector **p** – **q**. [2]

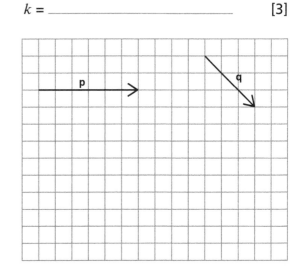

GCSE Maths Workbook

Total Marks: _____ / 12

GCSE
Mathematics
Paper 1 Foundation Tier

F

Materials

Time allowed: 1 hour 30 minutes

For this paper you must have:

- mathematical instruments

You may **not** use a calculator.

Instructions

- Use black ink or black ball-point pen. Draw diagrams in pencil.
- Answer **all** questions.
- You must answer the questions in the space provided.
- In all calculations, show clearly how you work out your answer.

Information

- The marks for questions are shown in brackets.
- The maximum mark for this paper is 80.
- You may use additional paper, graph paper and tracing paper.

Name: _____

Practice Exam Paper 1

Answer all questions in the spaces provided

1   Calculate the value of $9 - 2 \times 6 + 1$

Circle your answer.   **[1 mark]**

$$-5 \qquad -2 \qquad 43 \qquad 49$$

2   Work out the size of angle x.

Circle your answer.   **[1 mark]**

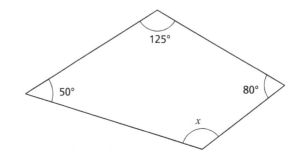

55 degrees   105 degrees   125 degrees   255 degrees

3   Work out $2^3 \times 3^2$

Circle your answer.   **[1 mark]**

$$17 \qquad 30 \qquad 36 \qquad 72$$

4   The perimeter of a square is 48 cm.

What is the area of the square?

Circle your answer.   **[1 mark]**

24 cm²   48 cm²   96 cm²   144 cm²

5   Simplify $2a + 3b - (a + b)$ **[2 marks]**

_____

_____

Answer _____

6   **(a)** Solve $6x - 2 = -5$ **[2 marks]**

_____

_____

$x = $ _____

(b) Solve $\frac{x}{3} = 12$ **[1 mark]**

_____

_____

$x = $ _____

7   Which of these is a multiple of 11?

Circle your answer.   **[1 mark]**

91   111   121   131

TURN OVER FOR THE NEXT QUESTION

8   A discount ticket from Wimbledon to Putney costs £1.50
      A single ticket costs £1.70
      Every week Tom makes six discount journeys and four single journeys.

      Work out the total travel costs for Tom during the 4 weeks of February.   **[3 marks]**

£ _____

9   These four cards have a median of 16 and a range of 8.

| 18 | 12 | ? | 14 |

      Work out the value of the missing card.   **[2 marks]**

Answer _____

10   Work out $183.4 \div 7$ **[2 marks]**

Answer _____

11   A traffic light can be on red, amber or green.

The probability that the traffic light is on red is 0.4

The probability that the traffic light is on amber is 0.15

(a) Work out the probability that the traffic light is on red or amber.   **[1 mark]**

_____

_____

Answer _____

(b) Work out the probability that the traffic light is on green.   **[2 marks]**

_____

_____

Answer _____

12   Complete the Venn diagram.

ξ − {odd numbers between 8 and 30}

P = {9, 13, 19, 21}

Q = {square numbers}

[2 marks]

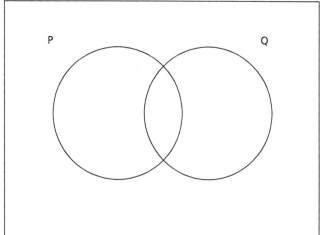

TURN OVER FOR THE NEXT QUESTION

13 (a) Work out $\frac{7}{8} + \frac{2}{3}$

Give your answer as a mixed number. **[3 marks]**

Answer _____

(b) Work out $\frac{4}{5}$ of 180 **[2 marks]**

Answer _____

14 (a) Complete the table of values below for the line with equation $y = \frac{1}{2}x - 1$ **[2 marks]**

x	−1	0	1	2	3
y		−1	$-\frac{1}{2}$		$\frac{1}{2}$

(b) Draw the line $y = \frac{1}{2}x - 1$ for values of x from −1 to 3 **[2 marks]**

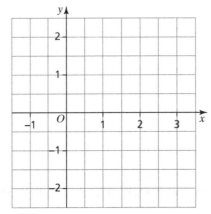

15 Solve the inequality $16 - \frac{x}{2} > 20$ [2 marks]

_____

_____

Answer _____

16 What is 0.091 written in standard form?

Circle your answer. [1 mark]

9.1×10^2 $\qquad$ 9.1×10^{-2} $\qquad$ 91×10^{-1} $\qquad$ 91×10^{-2}

17 In 2019 a school had 1200 students.
In 2020 the number of students increased by 15%.

(a) Work out the number of students in 2020. [2 marks]

_____

_____

Answer _____

(b) From 2020 to 2021, the number of students decreased by one-tenth.
Work out the number of students in 2021. [2 marks]

_____

_____

Answer _____

TURN OVER FOR THE NEXT QUESTION

18 Shop A sells a pack of five avocados for £1.40

Shop B sells a pack of four avocados for £1.16

Which shop gives the better value for money?
You **must** show all of your working. **[3 marks]**

Answer _____

19 Tiles are touching in a row on a kitchen wall as shown.

The wall is 1.4 m wide. The tiles are 250 mm wide.

(a) Work out the number of full tiles that can fit across the length of wall. **[2 marks]**

Answer _____

(b) Work out the width of the wall that would be left uncovered.
Give your answer in centimetres. **[2 marks]**

Answer _____ cm

20 (a) Work out the size of an exterior angle of a regular decagon. **[2 marks]**

_____

_____

_____

Answer _____ degrees

(b) Work out the size of an interior angle of a regular decagon. **[2 marks]**

_____

_____

_____

Answer _____ degrees

21 (a) What is the value of sin 30°?

Circle your answer. **[1 mark]**

-1 $-\dfrac{1}{2}$ 0 $\dfrac{1}{2}$ 1

(b) Work out the value of $\sin 45° - \cos 45°$

Circle your answer. **[1 mark]**

-1 $-\dfrac{1}{2}$ 0 $\dfrac{1}{2}$ 1

TURN OVER FOR THE NEXT QUESTION

22 Ali, Bob and Carla share some money in the ratio 4 : 6 : 11
Carla receives £200 more than Bob.

How much do they each receive? **[3 marks]**

_____

_____

_____

_____

Ali   £ _____   Bob   £ _____   Carla   £ _____

23 1 kilogram is approximately equal to 2.2 pounds.
A cake recipe requires 400 grams of flour.

How much would the flour weigh in pounds? **[3 marks]**

_____

_____

_____

Answer _____ pounds

24 Tim and Ann investigate the probability that it will rain on any given day in October.

Tim records the number of days that it rains in a week.

Ann records the number of days that it rains in the whole month.

Here are the results.

	Number of days it rains	Number of days without rain
Tim	2	5
Ann	11	20

(a) Write down **two** different estimates for the probability that it will rain on any given day in October. [2 marks]

_____

_____

Answer _____

(b) Which is the most reliable estimate from your answers in part **(a)**?

Give a reason for your answer. [2 marks]

Answer _____

Reason _____

_____

25 A man bought a house three years ago.

He sold it this year for £180 000, making 20% profit.

How much did he buy the house for three years ago? [3 marks]

_____

_____

£ _____

TURN OVER FOR THE NEXT QUESTION

26 y is directly proportional to x.

When $x = 2$, $y = 3$

Work out the value of y when $x = 16$ **[3 marks]**

_____

_____

_____

_____

$y = $ _____

27 Factorise $x^2 + 14x + 24$ **[2 marks]**

_____

_____

_____

_____

Answer _____

28 Solve the simultaneous equations

$$2x + 3y = 3$$

$$x - 5y = 8$$

[3 marks]

_____

_____

_____

_____

_____

$x =$ _____ $y =$ _____

29 A line has gradient 4 and goes through the point (–4, –9).

Work out the equation of the line.
Give your answer in the form $y = mx + c$

[3 marks]

_____

_____

_____

_____

_____

Answer _____

TURN OVER FOR THE NEXT QUESTION

30 The trapezium and the rectangle have the same area.

$x : y = 4 : 1$

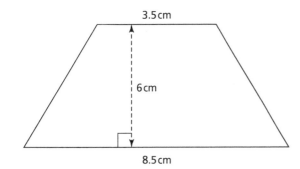

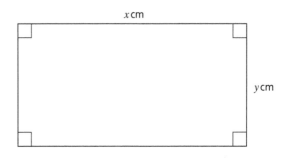

Work out the value of x and the value of y. **[5 marks]**

_____

_____

_____

_____

_____

_____

$x =$ _____ $y =$ _____

END OF QUESTIONS

Collins

GCSE
Mathematics
Paper 2 Foundation Tier

F

Materials

Time allowed: 1 hour 30 minutes

For this paper you must have:

- a calculator
- mathematical instruments.

Instructions

- Use black ink or black ball-point pen. Draw diagrams in pencil.
- Answer **all** questions.
- You must answer the questions in the space provided.
- In all calculations, show clearly how you work out your answer.

Information

- The marks for questions are shown in brackets.
- The maximum mark for this paper is 80.
- You may use additional paper, graph paper and tracing paper.

Name: _____

Practice Exam Paper 2

Answer all questions in the spaces provided

1 Which is the smallest number in the following list?

Circle your answer. **[1 mark]**

$$0.019 \qquad 0.01 \qquad 0.091 \qquad 0.09$$

2 Simplify $7 - 2x - 3 + x$

Circle your answer. **[1 mark]**

$$4 - x \qquad 4 - 3x \qquad 10 - x \qquad 10 - 3x$$

3 What is the probability of rolling a 3 on an ordinary fair dice?

Circle your answer. **[1 mark]**

$$\frac{1}{3} \qquad \frac{1}{2} \qquad \frac{1}{6} \qquad \frac{5}{6}$$

4 Which of these numbers is closest to the value 2?

Circle your answer. **[1 mark]**

$$\frac{\pi}{2} \qquad 1.2^3 \qquad \frac{18}{11} \qquad \frac{2}{3} \text{ of } 4$$

5 **(a)** Show the inequality $x > -2$ on this number line. [2 marks]

$$\begin{array}{ccccccccc} \vdash & + & + & + & + & + & + & + & \rightarrow \\ -4 & -3 & -2 & -1 & 0 & 1 & 2 & 3 & 4 \end{array} \; x$$

(b) Solve the inequality $7x + 1 < 29$ [2 marks]

_____

_____

Answer _____

6 The table shows how Mark spends his time on one day.

Activity	Sleeping	Eating	Homework	Relaxing	Attending school
Number of hours	9	$1\frac{1}{2}$	3	$4\frac{1}{2}$	6

Complete the pictogram below to illustrate the information shown in the table.

[3 marks]

Key: $\bigcirc$ = 3 hours

Activity	Number of hours
Sleeping	
Eating	
Homework	
Relaxing	
Attending school	

TURN OVER FOR THE NEXT QUESTION

7 Circle the highest common factor (HCF) of 8 and 12. **[1 mark]**

1   2   4   8

8 Lucy starts work each day at 8.45 am and finishes at 5.20 pm.

She takes 1 hour off for lunch.
She has two breaks of 20 minutes each.

How long does she work each day?

Circle your answer. **[1 mark]**

6 hours 25 minutes   6 hours 55 minutes

7 hours 35 minutes   7 hours 55 minutes

9 Alan has x buttons.

Barry has 10 more buttons than Alan.
Carl has twice as many buttons as Barry.

Work out an expression, in terms of x, for the total number of buttons.
Simplify your answer as much as possible. **[3 marks]**

_____

_____

_____

_____

Answer _____

10 A sequence has the position-to-term rule 7 – 5*n*

(a) Write down the first **three** terms in the sequence. [2 marks]

_____

_____

Answer _____

(b) Is –57 a term in this sequence?

Give a reason for your answer. [2 marks]

_____

_____

_____

_____

11 Rotate the triangle through an angle of 90° anticlockwise, with a centre of rotation (0, 0).

[2 marks]

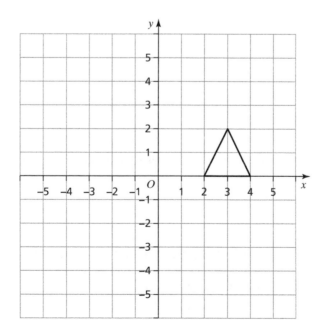

TURN OVER FOR THE NEXT QUESTION

12   On the map, assume any distances are straight lines.

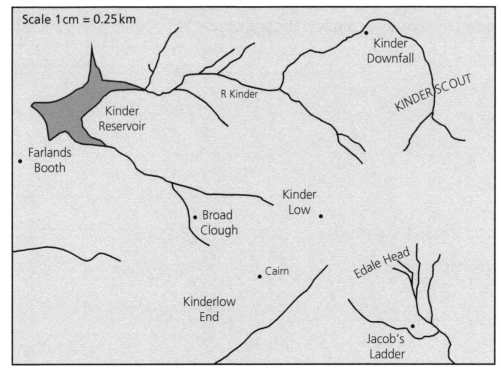

Lee walks from Kinder Downfall to Kinder Low.

He then walks from Kinder Low to Jacob's Ladder.

(a) Work out the actual distance he walks.   [2 marks]

_____

_____

Answer _____ km

(b) In fact, the distances are not straight lines. How would this affect the answer to part **(a)**?

[1 mark]

_____

13   The width of a desk, w, is measured as 183.5 cm to the nearest mm.

Circle the correct error interval for the width of the desk.   [1 mark]

$183.0 \text{ cm} \leqslant w < 184.0 \text{ cm}$ $183.49 \text{ cm} \leqslant w < 183.51 \text{ cm}$

$183.45 \text{ cm} \leqslant w < 183.55 \text{ cm}$ $183.4 \text{ cm} \leqslant w < 184.6 \text{ cm}$

14 **(a)** By approximating each value given to 1 significant figure, estimate the value of

$$\frac{1.79^2 + \sqrt{3.81}}{5.03 + 1.2}$$

Show your working.   **[2 marks]**

_____

_____

_____

_____

Answer _____

(b) State whether your answer is an underestimate or an overestimate.

Give a reason for your answer.   **[1 mark]**

_____

_____

Answer _____

15 Tins of beans are sold in boxes of 14.
Tins of soup are sold in boxes of 8.

Stu wants the same number of tins of beans and tins of soup.

How many boxes of beans and soup should he buy so that he has none left over?   **[3 marks]**

_____

_____

_____

_____

Boxes of beans _____   Boxes of soup _____

TURN OVER FOR THE NEXT QUESTION

16 The pie chart shows the proportions of the five most popular breeds of dogs in the UK in 2019.

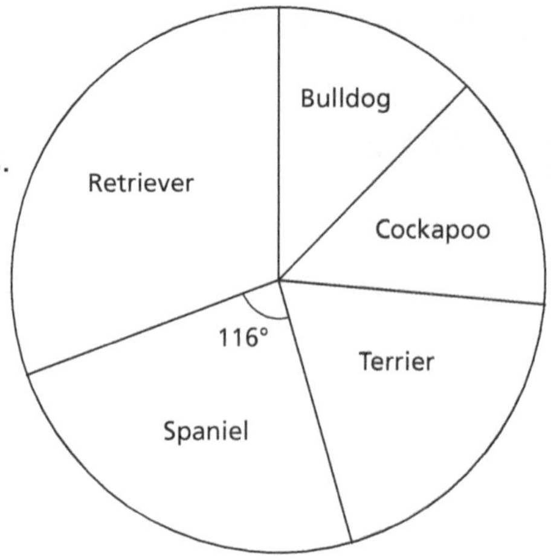

(a) Work out the percentage of dogs that are spaniels.

[2 marks]

_____

_____

Answer _____ %

(b) In 2020, the five most popular dog breeds remained the same.
The angle of the sector for spaniels increased from 116° to 120°.

Carla says, "The number of spaniels increased from 2019 to 2020."

Give **one** reason why Carla may be incorrect.

[1 mark]

_____

_____

17 Rachel (R), Jo (J) and Anuja (A) are competing in a running race.

Assume that there are no joint places and that each runner has an equal chance of winning.

(a) List all the different orders in which they could finish the race.

The first one has been done for you.

[2 marks]

R J A _____

_____

(b) Work out the probability that Jo will finish in a slower time than Rachel.

[2 marks]

_____

_____

Answer _____

18 Work out the value of $12 - 5\cos 36°$
Give your answer to 3 significant figures. **[1 mark]**

_____

Answer _____

19 Circle the value of 6×3^{-2} **[1 mark]**

$\dfrac{1}{81}$ $\dfrac{2}{3}$ $\dfrac{1}{324}$ $\dfrac{1}{9}$

20 A scientific formula for pressure (P) is given by $P = hdg$

These values are known: $d = 997$ $g = 9.8$ $P = 20\,000$

Work out the value of h.
Give your answer to 3 significant figures. **[3 marks]**

_____

_____

_____

$h = $ _____

TURN OVER FOR THE NEXT QUESTION

21

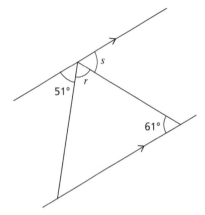

(a) Write down the size of angle *s*.
Give a reason for your answer.

[2 marks]

Answer _____ degrees

Reason _____

(b) Work out the size of angle *r*.

[1 mark]

_____

Answer _____ degrees

22 Here are two offers for the same shampoo.

Superstrands	**Hair-Care**
Buy 1 bottle for £3.59	Buy 1 bottle for £3.79
Buy two, get one free	40% off your total bill

Lynn wants three bottles.

Which shop gives the better value for money?
You **must** show your working.

[4 marks]

_____

_____

_____

Answer _____

23 Dan is making a concrete mix.
He mixes together 500 g of cement, 1 kg of sand and 1.5 kg of water.
He makes 2.17 litres of concrete.

Work out the density of the concrete.
Give your answer to 2 decimal places. **[2 marks]**

_____

_____

_____

Answer _____ g/cm³

24 This diagram shows a solid made from cubes.

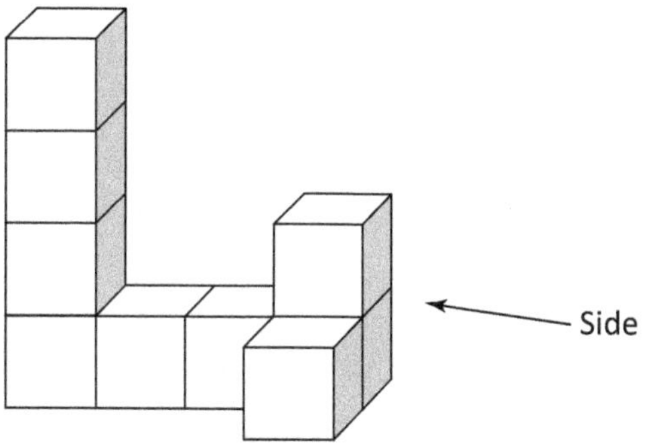

(a) Sketch the side elevation of the shape. **[2 marks]**

← Side

(b) Each cube has a side of length 1 cm.

Write down the volume of the solid. **[1 mark]**

Answer _____

TURN OVER FOR THE NEXT QUESTION

25 Mel changes £450 into euros (€) at an exchange rate of £1 = €1.285

(a) Work out the number of euros she should get. [2 marks]

_____

Answer € _____

(b) A bottle of perfume costs €36 in France or £27 in the UK.

Work out the difference in price between the two countries.
Give your answer in pounds. [2 marks]

_____

_____

Answer £ _____

26 The table shows the time taken for 20 runners to complete a 100 m race.

Time, t seconds	Number of runners	Midpoint	
$12 \leqslant t < 13$	1		
$13 \leqslant t < 14$	5		
$14 \leqslant t < 15$	11		
$15 \leqslant t < 16$	3		

(a) Complete the table and calculate an estimate for the mean time. [3 marks]

_____

Answer _____ seconds

(b) Work out the largest possible range of times from the information in the table. [1 mark]

_____

Answer _____ seconds

27 There are 162 students in Year 10.

$\frac{2}{3}$ of the students go to a theme park.

$\frac{8}{27}$ of the students go ice-skating.

The rest of the students stay in school.

How many of the students stay in school?   **[4 marks]**

_____

_____

_____

_____

Answer _____ students

28 Sophia cuts a length of string into three smaller pieces in the ratio 3 : 2 : 7
The smallest piece of string is 15 cm.

How long was the original length of string?   **[2 marks]**

_____

_____

Answer _____ cm

29 Ahmed earns £56 000 per year.
The first £12 000 he earns is not taxed.
The next £38 000 he earns is taxed at 20%
The remainder is taxed at 40%

Calculate the amount he earns after the tax has been deducted.   **[3 marks]**

_____

_____

Answer £ _____

TURN OVER FOR THE NEXT QUESTION

30 The composite shape *ADCB* shows a sector of a large circle joined to a sector of a smaller circle.

Work out the total area of the shape.
Give your answer to 3 significant figures.

[4 marks]

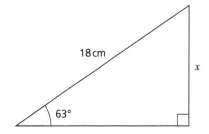

_____

_____

_____

_____

Answer _____ cm²

31 Here is a right-angled triangle.

18 cm

x

63°

Work out the length x.
Give your answer to 1 decimal place.

[3 marks]

_____

_____

_____

Answer _____ cm

END OF QUESTIONS

Collins

GCSE
Mathematics
Paper 3 Foundation Tier

F

Materials

Time allowed: 1 hour 30 minutes

For this paper you must have:
- a calculator
- mathematical instruments.

Instructions

- Use black ink or black ball-point pen. Draw diagrams in pencil.
- Answer **all** questions.
- You must answer the questions in the space provided.
- In all calculations, show clearly how you work out your answer.

Information

- The marks for questions are shown in brackets.
- The maximum mark for this paper is 80.
- You may use additional paper, graph paper and tracing paper.

Name: _____

Practice Exam Paper 3

Answer all questions in the spaces provided

1 Which of these is the greatest length?

Circle your answer. **[1 mark]**

150 mm   0.001 km   150 cm   0.15 m

2 What is the lowest common multiple of 15 and 20?

Circle your answer. **[1 mark]**

5   35   60   300

3 Convert 3.5 kg into grams.

Circle your answer. **[1 mark]**

0.35 grams   35 grams   350 grams   3500 grams

4 A fair, six-sided dice is thrown.

What is the probability of throwing a prime number?

Circle your answer. **[1 mark]**

0          $\frac{1}{2}$ $\frac{1}{3}$ $\frac{2}{3}$

5 A shape has these properties

- it has four sides
- all the sides are equal in length
- opposite angles are equal
- none of its interior angles are equal to 90 degrees

Make a sketch of the shape.   **[2 marks]**

6 The total mass of 12 identical washing machines is 1625.22 kg.

(a) Work out the total mass of 15 of these machines.
Give your answer to 3 decimal places.   **[2 marks]**

_____

_____

Answer _____ kg

(b) The maximum load of a lorry is 2.6 tonnes.

Work out the greatest number of these washing machines it can carry.   **[2 marks]**

_____

_____

Answer _____

TURN OVER FOR THE NEXT QUESTION

7 Work out the median of this set of numbers.

2   5   10   13   13   20   **[1 mark]**

_____

Answer _____

8 Here is a number machine.

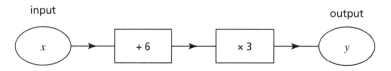

(a) Work out the output when the input is 10   **[1 mark]**

_____

Answer _____

(b) Work out the input when the output is 27   **[2 marks]**

_____

Answer _____

(c) Circle a correct expression for y in terms of x.   **[1 mark]**

$y = x + 18$ $y = 3x + 6$ $3y + 6 = x$ $y = 3(x + 6)$

9 Work out $a : c$ given that $a : b = 2 : 5$ and $b : c = 1 : 4$
Circle your answer.   **[1 mark]**

1 : 2   1 : 4   1 : 8   1 : 10

10   Circle the inequality for the set of integers   −1, 0, 1, 2, 3, 4   **[1 mark]**

$-1 \leqslant x \leqslant 5$ $-1 < x < 5$ $-1 \leqslant x < 5$ $-1 < x \leqslant 5$

11   The total cost (£C) to learn to drive is given by

> The number of lessons, n, multiplied by £25
>
> plus
>
> The number of tests, t, multiplied by £85

(a) Write a formula for C in terms of n and t.   **[2 marks]**

_____

Answer _____

(b) Lorna has 17 lessons and passes the test on her second attempt.

Work out the total cost for Lorna to learn to drive.   **[2 marks]**

_____

_____

Answer £ _____

(c) Vik passed his test on the first attempt.
The total cost was £560

How many lessons did Vik have?   **[2 marks]**

_____

Answer _____

TURN OVER FOR THE NEXT QUESTION

12 An internet company said that monthly broadband usage had increased from 270 GB to 402.5 GB per customer.

Work out the percentage increase. **[2 marks]**

_____

_____

Answer _____ %

13 $(a + 2)x - 3b \equiv 5x + 9$

Which one of the following is true?

Circle your answer. **[1 mark]**

$a = 2$ and $b = 3$ $a = 2$ and $b = -3$ $a = 3$ and $b = 3$ $a = 3$ and $b = -3$

14 Two sides of a regular hexagon are extended to form a triangle as shown.

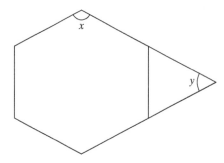

(a) Work out the size of angle x. **[2 marks]**

_____

$x = $ _____ degrees

(b) Work out the size of angle y. **[2 marks]**

_____

$y = $ _____ degrees

15 Carl can run 100 m in 10 seconds.

Work out his speed in kilometres per hour. **[3 marks]**

_____

_____

_____

Answer _____ km/h

16 **(a)** Work out 8600 × 45

Give your answer in standard form. **[2 marks]**

_____

_____

_____

Answer _____

(b) Work out 1.6 × 10^5 ÷ 500

Give your answer in standard form. **[2 marks]**

_____

_____

_____

Answer _____

TURN OVER FOR THE NEXT QUESTION

17   A fuel tank when empty holds 60 litres.

Theo spends £35.60 to fill the tank.
The fuel costs £1.28 per litre.

Work out the number of litres of fuel that was in the tank before it was filled up.
Give your answer to the nearest litre.   **[4 marks]**

_____

_____

_____

_____

Answer _____ litres

18   **(a)** Factorise fully $9x^2y - 18xy^2$ **[2 marks]**

_____

_____

_____

Answer _____

(b) Factorise $x^2 - 16$ **[2 marks]**

_____

_____

_____

Answer _____

19 Here is a sequence.

$$2 \qquad 4 \qquad 6 \qquad 10 \qquad \ldots \qquad \ldots$$

(a) Work out the next **two** terms. [2 marks]

_____

_____

Answer _____

(b) Here is an arithmetic sequence.

$$-2 \qquad 6 \qquad 14 \qquad 22 \qquad \ldots \qquad \ldots$$

Work out the rule for the nth term of this sequence. [2 marks]

_____

_____

Answer _____

20 The time spent by Ali and Sue on a task is in the ratio 5 : 2
Ali spends 9 hours more on the task than Sue.

How many hours did Sue spend on the task? [3 marks]

_____

_____

_____

Answer _____ hours

TURN OVER FOR THE NEXT QUESTION

21   The tables show the salary of ten midwives and the number of years worked.

Years worked	2	5	3	9	8
Salary	£22 000	£28 000	£24 000	£29 500	£30 000

Years worked	3	4	2	7	5
Salary	£23 000	£27 000	£24 000	£25 500	£26 000

(a) Complete the scatter graph to represent this information.
The first five have been done for you.   **[2 marks]**

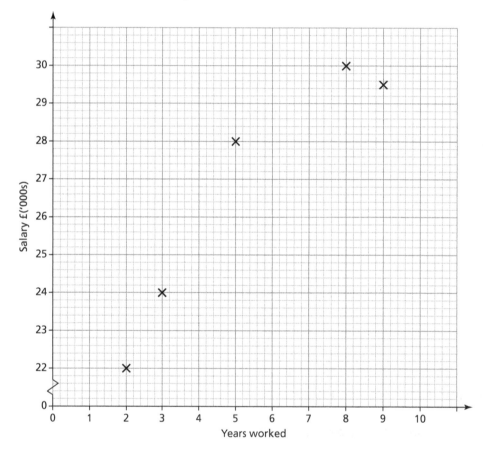

(b) Estimate the salary of a midwife who has been working for 6 years.   **[1 mark]**

_____

Answer £ _____

22 Three positive whole numbers have a total of 50
The first number is a two-digit square number.
The second and third numbers are in the ratio 1 : 4

Work out the three numbers. [3 marks]

_____

_____

Answer _____, _____, _____

23 Write 15 125 as a product of prime factors.
Give your answer in index form. [3 marks]

_____

_____

Answer _____

24 A school estimates that $\frac{1}{8}$ of its students do not bring a calculator for an exam.

There are 224 students taking the exam.

(a) Estimate the number of students who do not bring a calculator. [2 marks]

_____

Answer _____

(b) 215 of the students pass the exam.

Work out the percentage of the students that pass the exam. [2 marks]

_____

Answer _____ %

25 Asim goes to the gym every Monday and Tuesday.

The probability that he walks to the gym each day is 0.35

(a) Complete the tree diagram. **[2 marks]**

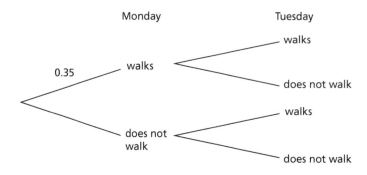

(b) Work out the probability that Asim does not walk to the gym on either day. **[2 marks]**

Answer _____

26 $\mathbf{a} = \begin{pmatrix} 4 \\ -3 \end{pmatrix}$ and $\mathbf{b} = \begin{pmatrix} -1 \\ 2 \end{pmatrix}$

Bel says, "The vector $\begin{pmatrix} -10 \\ 10 \end{pmatrix}$ is parallel to $\mathbf{a} - \mathbf{b}$."

Is she correct?

Give a reason for your answer. **[2 marks]**

27 Here are six graphs.

A   B   C   D   E   F

Write down the letter of the graph which matches these equations.

(a) $y = 3x - 2$ Graph _____ **[1 mark]**

(b) $y = x^2$ Graph _____ **[1 mark]**

(c) $y = \dfrac{1}{x}$ Graph _____ **[1 mark]**

28   The scale drawing shows a boat (B) and some rocks (R).

Scale 1 cm : 100 m

R
×

N

B

(a) How far is the boat from the rocks?   [2 marks]

_____

_____

Answer _____ m

(b) Work out the bearing of the rocks from the boat.   [2 marks]

_____

_____

Answer _____ °

(c) A ship is going to pass between the boat and the rocks.

It must pass closer to the boat than the rocks.

Shade the region through which the ship can pass.   [2 marks]

END OF QUESTIONS

Answers

Workbook Answers

You are encouraged to show all your working out, as you may be awarded marks for method even if your final answer is wrong. Full marks can be awarded where a correct answer is given without working but, if a question asks for working, you must show it to gain full marks. If you use a correct method that is not shown in the answers below, you would still gain full credit for it.

Page 4: Number 1, 2 & 3

1. $3 + 12 - 6 = 9$ [1]
2. $\frac{1}{6} + \frac{4}{6} = \frac{5}{6}$ [1]
3. 0.0045 [1]
4. 1817 [1]
5. a) 37 [1]
   b) 36 [1]
   c) 30 [1]
   d) 24 [1]
   e) 14 [1]
6. a) −15 [1]
   b) 15 [1]
   c) 36 [1]
7. $6 \times £6.50 = £39$ [1]
   $£50 - £39 = £11$ [1]
8. $3 \times 18 + 6$ [1]
   $= 60$ [1]
9. $3600 - 110$
   $= 3490$ [1]
   $= 3.49 \times 10^3$ [1]
10. $147\,000\,000\,000$ [1]
    $= 1.47 \times 10^{11}$ m [1]
11. $x = 12$ [1] or $x = 24$ [1]
12. $84 = 2 \times 2 \times 3 \times 7$ [2]
    [1 mark if 2 ×... seen]
13. 25×10^{10} [1]
    $= 2.5 \times 10^{11}$ [1]
14. $\left(-\frac{5}{2}\right)^3$ [1]
    $= -\frac{125}{8}$ [1]

Page 6: Basic Algebra

1. $x = 27$ [1]
2. $8x = 32$ [1]
   $x = 4$ [1]
3. $3y - 8x$ [1]
4. $9p - 3q$ [1]
5. $14y - 21 + 12y + 8$ [1]
   $= 26y - 13$ [1]
6. $4 + 25$ [1]
   $= 29$ [1]
7. $6 - 2(2 - x) = 20$
   $6 - 4 + 2x = 20$ [1]

$2x = 18$
$x = 9$ [1]
8. $3x - 6 = 84 - 12x$ [1]
   $15x = 90$ [1]
   $x = 6$ [1]

Page 7: Factorisation and Formulae

1. $9q(3p^3 - 2q)$ [1]
2. $x^2 + 10x + 2x + 20$ [1]
   $= x^2 + 12x + 20$ [1]
3. $(x + 6)(x - 6)$ [1]
4. a) $150 + 66 = £216$ [1]
   b) $C - 150 = 12t$ [1]
      $t = \frac{C - 150}{12}$ [1]
   c) $\frac{350 - 150}{12}$ [1]
      $= 16.7$ [1]
      So 16 hours [1]
5. $(x + 7)(x - 6)$ [2]
   [1 mark for each correct bracket]

   > Remember you are looking for two numbers that add to give 1 and multiply to give −42.

6. $5x - 10 = 2q + 10$ [1]
   $5x = 2q + 20$ [1]
   $x = \frac{2q + 20}{5}\left(= \frac{2q}{5} + 4\right)$ [1]

Page 8: Ratio and Proportion

1. $1 : \frac{1}{2}$ [1]
2. Shop A: one bar costs 62.5p
   Shop B: one bar costs 63.9p [1]
   Shop A offers the better value [1]
3. $\frac{£360}{8} = £45$ [1]
   $3 \times £45 = £135$ [1]
   $5 \times £45 = £225$ [1]
4. $3 : 2 : 10$ [1]
5. $1 : 1.25$ or $1 : \frac{5}{4}$ [1]
6. $12 \times \frac{8}{5}$ [1]
   $= 19.2\,l$ [1]
7. 57 km $= 5\,700\,000$ cm [1]
   $5\,700\,000 \div 250\,000$ [1]
   $= 22.8$ cm [1]

Page 9: Variation and Compound Measures

1. y is doubled [1]

   > y is directly proportional to x means $y = kx$, so if x is doubled (the right-hand side), then so is the left-hand side.

2. $70 \div 2.237$ [1]
   $= 31.3$ m/s [1]

3. $2500 \div 45 = 55.6 \text{ N/m}^2$ [1]

Pressure = Force ÷ Area

4. $0.15 \text{ kg} = 150 \text{ g}$ [1]

$150 \div 7.13$ [1]

$= 21.0 \text{ cm}^3$ [1]

Density = Mass ÷ Volume

5. a) $k = 2$ [1]

b) $25 = 2x$ [1]

$x = \frac{25}{2} = 12.5$ [1]

6. a) $k = 20$ [1]

b) $y = 20 \div \frac{1}{2}$ [1]

$= 40$ [1]

Page 10: Angles and Shapes 1 & 2

1. $3x + x + 118 + 114 = 360$ [1]

$4x + 232 = 360$

$4x = 128$ [1]

$x = 32°$ [1]

2. a) $3 \times 180 = 540°$ [1]

A pentagon can be split into three triangles.

b) $x = \frac{540}{5} = 108°$ [1]

Considering quadrilateral *ABCE*:

$y + y + 108 + 108 = 360$ [1]

$y = 72°$ [1]

y may also be found by considering the isosceles triangle *CDE*.

3. Isosceles triangle base angle =

$(180 - 38) \div 2 = 71°$ [1]

Vertically opposite angle = 71° [1]

$x = 180 - 90 - 71 = 19°$ [1]

4. 180×8 **[1]** $= 1440°$ **[1]**

Page 11: Fractions

1. $70 \div 5 = 14$

$14 \times 2 = £28$ [1]

2. Using denominator of 32:

$\frac{24}{32}, \frac{21}{32}, \frac{20}{32}, \frac{26}{32}, \frac{16}{32}$ [1]

So $\frac{1}{2}, \frac{5}{8}, \frac{21}{32}, \frac{3}{4}, \frac{13}{16}$ [1]

[1 mark if three in the correct order]

3. $\frac{1500 - 350}{1500}$ [1]

$= \frac{1150}{1500}$

$= \frac{23}{30}$ [1]

4. 60 chocolates $= \frac{5}{6}$ box [1]

1 box $= 60 \times \frac{6}{5} = 72$ chocolates [1]

5. $\frac{1}{2} \times \frac{3}{4} \times 64$ [1]

$= \frac{3}{8} \times 64$

$= 24$ [1]

6. $\frac{13}{4} \div \frac{7}{5}$ [1]

$\frac{13}{4} \times \frac{5}{7}$ [1]

$= \frac{65}{28}$ [1]

$= 2\frac{9}{28}$ [1]

Page 12: Percentages 1 & 2

1. $\frac{75 - 32}{32} \times 100$ [1]

$= \frac{43}{32} \times 100$

$= 134.4\%$ [1]

2. $1.5 \, l = 1500 \text{ ml}$ [1]

$\frac{900}{1500} \times 100$ [1]

$= 60\%$ [1]

3. a) $24\,000 \div 12 = £2000$ [1]

b) $2000 - \frac{18}{100} \times 2000$ [1]

$= £1640$ [1]

4. 125% of $1400 = \frac{125}{100} \times 1400$ [1]

$= £1750$ [1]

$£1400 + £1750 = £3150$ [1]

5. 5% of $x = 1995$

$x = \frac{1995}{0.05}$ [1]

$= 39\,900$ [1]

Page 13: Probability 1 & 2

1. a) $\frac{1}{2}$ [1]

b) 25% [1]

c) 0.6 [1]

2. a) $\frac{11}{20}$ [1]

b) $\frac{13}{20}$ [1]

c) 0 [1]

3. a)

		2nd spin		
	2	3	4	6
2	4	5	6	8
3	5	6	7	9
4	6	7	8	10
6	8	9	10	12

[2]

(1st spin labels rows 2, 3, 4, 6)

[1 mark lost for each error up to a maximum of 2]

b) $\frac{2}{16}\left(= \frac{1}{8}\right)$ [1]

c) $\frac{13}{16}$ [1]

4. a) 0.2   **[1]**

b)

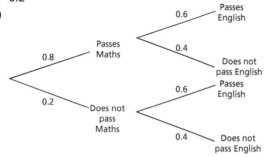

First column of probabilities correct   **[1]**
Second column of probabilities correct   **[1]**

c) $(0.8 \times 0.4) + (0.2 \times 0.6)$ **[2]**
$= 0.32 + 0.12$
$= 0.44$ **[1]**

5. a) If the coin was fair (i.e. not biased), the probability
of throwing a head would be 0.5   **[1]**

b) P(Tails) = 0.42   **[1]**
$0.42 \times 250 = 105$ **[1]**

Page 15: Number Patterns and Sequences 1

1. 10 **[1]** and 210 **[1]**

2. a) 17 **[1]**; 21 **[1]**

b) 15 **[1]**; 21 **[1]**

c) 64 **[1]**; 125 **[1]**

3. a) 23 and 26   **[1]**

b) $3n + 8$ **[2]**
[1 mark for $3n$ and 1 mark for 8]

c) 503   **[1]**

4. a) 29 **[1]** and 46 **[1]**

b) $\frac{8}{27}$ **[1]** and $\frac{16}{81}$ **[1]**

5. $6n + 3$ **[2]**
[1 mark for $6n$ and 1 mark for 3]

Page 16: Number Patterns and Sequences 2

1. a)

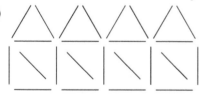

   [1]

b)

Pattern number	1	2	3	4
Number of sticks	7	13	19	25

   [2]
[1 mark for 19 and 1 mark for 25]

c) $6n + 1$ **[2]**
[1 mark for $6n$ and 1 mark for +1]

d) $6n + 1 = 150$ **[1]**
$n = \frac{149}{6} = 24.83\ldots$, not a whole number   **[1]**
So not possible.   **[1]**

2. a) 1, −2, −5, −8   **[2]**
[1 mark lost for each error up to a maximum of 2]

b) −296   **[1]**

Page 17: Transformations

1. a)

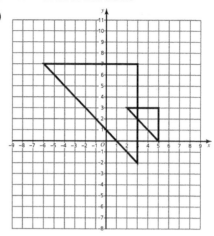

Fully correct   **[2]**
[1 mark if one vertex incorrectly plotted]

b) (5, 3), (5, 0) and (2, 3)   **[1]**

2. a) Rotation **[1]**; 90° anticlockwise **[1]**; about (0, 0) **[1]**

b) Reflection **[1]**; in x-axis (or in line $y = 0$) **[1]**

c) Sunil is correct. The three triangles are identical,
since they have not changed size. **[1]**

Page 18: Constructions

1.

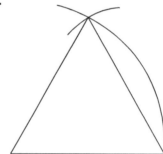

Arc drawn from one end with radius equal to
length of line   **[1]**
Equal arc from other end of line   **[1]**
Equilateral triangle drawn   **[1]**

2.

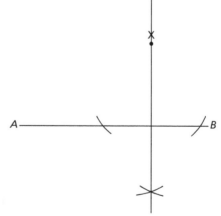

Arcs marked on line AB **[1]**
Remaining arcs marked   **[1]**
Straight vertical line drawn through arc   **[1]**

3.

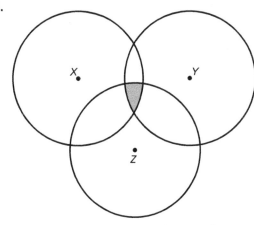

Three circles drawn of equal size, radius 2 cm [1]
Circles intersect to form seven separate regions [1]
Central region (only) shaded [1]

Page 19: Nets, Plans and Elevations

1.

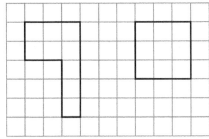

a) Fully correct (see above left) [1]
b) Fully correct (see above right) [1]
c) 27 [1]

2.

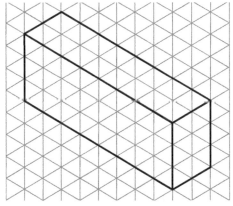

A cuboid with all lengths correct [3]
[1 mark lost for each incorrect length up to a maximum of 3]
(Allow the base of the cuboid drawn to be any of the three differing sides)

Page 20: Linear Graphs

1. a) Graph goes through (0, 3) [1]

$$m = -\frac{3}{6}\left(= -\frac{1}{2}\right)$$ [1]

Gradient = y step ÷ x step

b) When $x = -12$, $y = -\frac{1}{2}(-12) + 3 = 9$ [1]

So (–12, 10) is above the line. [1]

2. Gradients are: 1, 4, $\frac{1}{4}$, 4

So B and D are parallel [1]

Rearrange each line into the form $y = mx + c$

3. a) (–6, 0) [1]

b) $\left(0, \frac{6}{5}\right)$ [1]

4. $m = \frac{12}{6} = 2$ [1]

$y = 2x + c$

Substituting $x = 4$, $y = 2$ [1]

$2 = 8 + c$ [1]

$c = -6$ [1]

So $y = 2x - 6$

Page 21: Graphs of Quadratic Functions

1. a) (–2, 9) [1]

b) $x = 1$ [1]

$x = -5$ [1]

2.

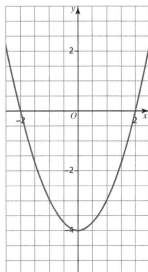

Crosses x-axis at –2 and 2 [1]
Crosses y-axis at –4 [1]
Correct shape [1]

Page 22: Powers, Roots and Indices

1. a) $2p^7$ [1]

b) q^6 [1]

c) $4r^6$ [1]

2. $20x^5y^3$ [2]

[1 mark for '20' and 1 mark for x^5y^3]

3. $2^{-1}, 3^0, 2^3, 3^2$ [2]

[1 mark for three in the correct order]

4. $9 - 4$ [1]

$= 5$ [1]

5. $3^{-2} = \frac{1}{3^2}$ [1]

$= \frac{1}{9}$ [1]

6. $\sqrt{12^2 + 5^2} = \sqrt{144 + 25} = \sqrt{169}$ [1]

$= 13$ [1]

Page 23: Area and Volume 1, 2 & 3

1. Area $= \frac{1}{2}(a+b)h$ [1]

     $= \frac{1}{2}(7.5 + 12.6) \times 6.4$ [1]

     $= 64.3$ m²   [1]

2. Arc length $= \frac{110}{360} \times 2\pi \times 20 (= 38.397)$ [2]

     Perimeter = 20 + 20 + arc length   [1]

     = 78.4 cm   [1]

3. $\pi r^2 = 50$ [1]

     $r \left(= \sqrt{\frac{50}{\pi}}\right) = 3.9894$ cm   [1]

     Circumference $= 2\pi r$ [1]

     $= 2\pi \times 3.9894 = 25.1$ cm   [1]

4. Area of rectangle $= 12 \times 8 = 96$ cm²   [1]

     Area of circles $= 6\pi r^2 = 6\pi \times 2^2 = 24\pi$ [2]

     Shaded area $= 96 - 24\pi = 20.6$ cm²   [1]

5. Volume of large cone $= \frac{1}{3}\pi r^2 h = \frac{1}{3} \times \pi \times 2^2 \times 6 = 8\pi$ [2]

     Radius of small cone = 1 cm   [1]

     Volume of small cone $= \frac{1}{3}\pi r^2 h = \frac{1}{3} \times \pi \times 1^2 \times 3 = \pi$ [2]

     Volume of frustum $= 8\pi - \pi = 7\pi$ cm³   [1]

6. Surface area of hemisphere $= \frac{4\pi r^2}{2} + \pi r^2$ [1]

     Substituting $r = 5$ [1]

     Surface area = 236 cm² (or 75π cm²)   [1]

> The formula for the curved surface area of a sphere is $4\pi r^2$.

Page 25: Uses of Graphs

1. a) $20\,000 - 1600 \times 3 = £15\,200$ [1]

     b) −1600   [1]

     c) The rate at which the car's value depreciates (per year)   [1]

     d) Car has no value when $20\,000 - 1600t = 0$ [1]

       $t = 12.5$ years   [1]

     e) Car is likely to lose value at a greater rate when it is new. Or

       Car is unlikely to become completely worthless.   [1]

2. A   [1]

3. D   [1]

Page 26: Other Graphs

1. a) 65 zloty   [1]

     b) 100 zloty = £19   [1]

       So 500 zloty = £95   [1]

2. a) 50 ÷ 1.5 **[1]** = 33.3 cm/h² **[1]**

     b) Travels at a constant speed of 50 cm/h   [1]

3.
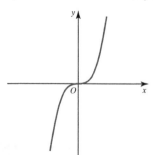

Graph goes through the point (0, 0)   [1]

Correct shape of graph   [1]

Page 27: Inequalities

1. a) $x < 18$ **[1]**   b) $x \geq 4$ **[1]**   c) $x \leq 7$ **[1]**

2. $x = -1, 0, 1, 2, 3$ [1]

3. $5x - 10 > 20$ [1]

     $5x > 30$

     $x > 6$ [1]

4. $-\frac{16}{5} \leq x < 4$ [1]

     $x = -3, -2, -1, 0, 1, 2, 3$ [1]

5. $-2 \leq x < 3$ [1]

6.

     Empty circle marked at −1   [1]

     Arrow to the right   [1]

7. $2x \leq 6$, so $x \leq 3$ [1]

     $x = 0, 1, 2, 3$ [2]

     [1 mark for three numbers correct]

Page 28: Congruence and Geometrical Problems

1. Scale factor $= \frac{27}{9} = 3$ [1]

     5 × 3 = 15 cm   [1]

2. A and C   [1]

     Two angles and a corresponding side are equal (AAS)   [1]

3. a) 60°   [1]

     b) $3 \times \frac{5}{2}$ [1]

       = 7.5 cm   [1]

     c) $9 \times \frac{2}{5}$ [1]

       = 3.6 cm   [1]

Page 29: Right-Angled Triangles 1

1. $PR^2 = 6.1^2 + 17.5^2$ [1]

     $PR = 18.5$ cm   [1]

2. $x^2 + 10^2 = 20^2$ [1]

     $x^2 + 100 = 400$ [1]

     $x^2 = 300$

     $x = 17.3$ cm   [1]

3. Length of diagonal $= \sqrt{1.2^2 + 0.7^2} = 1.389$ [2]

     Total length $= (0.7 \times 2) + (1.2 \times 4) + 1.389$ [1]

     = 7.59 m   [1]

4. C   [1]

     $20^2 + 23^2 \neq 31^2$ [1]

5. a) $h = \sqrt{20^2 - 13^2}$ [1]

       = 15.2 cm   [1]

     b) $\frac{1}{2} \times 26 \times 15.2$ **[1]** = 197.6 cm² **[1]**

Page 30: Right-Angled Triangles 2

1. $\cos 70 = \frac{x}{5.5}$ [1]

     $x = 5.5\cos 70 = 1.88$ m   [1]

2. $\tan 25 = \frac{x}{9}$ [1]

     $x = 9\tan 25 = 4.2(0)$ cm   [1]

3. a) $\sin x = \frac{9}{13}$ [1]

       $x = \sin^{-1}\left(\frac{9}{13}\right) = 43.8°$ [1]

b) $y = \cos^{-1}\left(\dfrac{9}{13}\right) = 46.2°$ **[1]**

   (or $y = 180 - 90 - 43.8 = 46.2°$)

4. Exactly three of them are true **[1]**

Page 31: Statistics 1

1. a) $28 \div 3.5$ **[1]** $= 8$ **[1]**

   Key: $\bigcirc$ = 8 **[1]**

b) 64 **[1]**

c)

9	$\bigcirc$ $\bigcirc$

Left-hand circle **[1]**

Right-hand three-quarter circle **[1]**

2. a) Rock **[1]**

b) 15×6 **[1]** $= 90$ students **[1]**

c) $\dfrac{108}{360} \times 90$ **[1]**

    $= 27$ students **[1]**

d) $\dfrac{24}{360} \times 100$ **[1]**

    $= 6.67\%$ **[1]**

Page 32: Statistics 2

1.

Mass of cat, M kg	Number of cats	Mid-point	Midpoint × Frequency
$3 \leqslant M < 4$	4	3.5	14
$4 \leqslant M < 5$	5	4.5	22.5
$5 \leqslant M < 6$	9	5.5	49.5
$6 \leqslant M < 7$	7	6.5	45.5

Correct table **[1]**

Mean mass $= \dfrac{14 + 22.5 + 49.5 + 45.5}{25} = \dfrac{131.5}{25}$ **[1]**

$= 5.26$ kg **[1]**

2. Let the mum's age be x:

$\dfrac{14 + 16 + 41 + x}{4} = 27$ **[1]**

$71 + x = 108$

$x = 37$ **[1]**

3. a) Negative correlation or

   As distance increases, price decreases **[1]**

b) Line of best fit drawn **[1]**

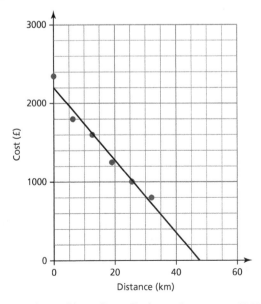

Use line of best fit to find cost is approx. £440 **[1]**

c) 38 km is outside the range of the distances given in the graph. Or

Line indicates rent would eventually become zero/ negative, so unrealistic. **[1]**

Page 33: Measures, Accuracy and Finance

1. a) 3.14 **[1]**

b) 3.142 **[1]**

c) 3.1416 **[1]**

d) 3.14159 **[1]**

2. $15.5 \times 12 \times 2.54 = 472.44$ cm **[1]**

    $472.44 \div 100 = 4.72$ m **[1]**

3. $\dfrac{9.09 \times 5.9}{(31 - 12.9) \times 3.1} \approx \dfrac{9 \times 6}{18 \times 3}$ **[1]**

    $\left(= \dfrac{54}{54}\right) = 1$ **[1]**

Start by rounding each number to 1 s.f.

4. Attempt to subtract 13 hours **[1]**

    8.30 pm (the previous evening) **[1]**

5. $\dfrac{15.26}{1.4}$ **[1]**

    $= £10.9(0)$ **[1]**

Page 34: Quadratic and Simultaneous Equations

1. Adding equations gives $10x = 10$ **[1]**

   So $x = 1$ **[1]**

    $y = 3$ **[1]**

2. a) $x = 2$ **[1]**

    $x = -3$ **[1]**

b) $x = 1$ **[1]**

    $x = -2$ **[1]**

3. a) $(x + 7)(x - 2)$ **[2]**

   [1 mark for each correct bracket]

b) $(x + 7)(x - 2) = 0$

    $x + 7 = 0$ or $x - 2 = 0$

    $x = -7$ or $x = 2$ **[1]**

Page 35: Circles

1. a) diameter **[1]**

b) tangent **[1]**

2. a) $\pi r^2 = \pi \times 4^2 = 16\pi$ cm^2 **[1]**

b) Fraction of the circle required is

    $\dfrac{2\pi}{16\pi} = \dfrac{1}{8}$ **[1]**

    $\dfrac{1}{8} \times 360 = 45°$ **[1]**

Sector drawn of internal angle 45° **[1]**

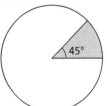

3. Area of quarter-circle

$= \dfrac{\pi r^2}{4} = \dfrac{\pi \times 15^2}{4} = 176.71\ldots$ **[2]**

Area of triangle $OAB = \frac{1}{2} \times 15 \times 15 = 112.5$ [2]

Shaded area $= 176.71... - 112.5 = 64.2\,cm^2$ [2]

Page 36: Vectors

1. a) Translation through $\begin{pmatrix} -1 \\ 2 \end{pmatrix}$ [2]

   [1 mark for each correct value]

   b) Translation through $\begin{pmatrix} 1 \\ -2 \end{pmatrix}$ [2]

   [1 mark for each correct value]

2. a) **a** and **c**. Two vectors are parallel if one is a multiple

   of the other, i.e. $\begin{pmatrix} 2 \\ -4 \end{pmatrix} = 2\begin{pmatrix} 1 \\ -2 \end{pmatrix}$ [1]

   b) $\begin{pmatrix} -6 \\ 1 \end{pmatrix} + \begin{pmatrix} 2 \\ -4 \end{pmatrix}$ [1]

   $= \begin{pmatrix} -4 \\ -3 \end{pmatrix}$ [1]

   c) $5\mathbf{a} + 2\mathbf{b} = \begin{pmatrix} 10 \\ -20 \end{pmatrix} + \begin{pmatrix} -12 \\ 2 \end{pmatrix} = \begin{pmatrix} -2 \\ -18 \end{pmatrix}$ [2]

   $k = -2$ [1]

3.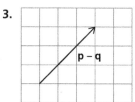

   Straight line crossing 'three diagonals' [1]
   Arrow pointing in direction shown [1]

Pages 37–50: Practice Exam Paper 1

1. -2 [1]
2. $105°$ [1]

   > The sum of the interior angles of a
   > quadrilateral add up to $360°$.

3. 72 [1]
4. $144\,cm^2$ [1]
5. $a + 2b$ [2]

   [1 mark for each correct term]

   > Collect the like terms.

6. a) $6x = -3$ [1]

   $x = -\frac{3}{6}$ (or $-\frac{1}{2}$ or -0.5) [1]

   b) $x = 36$ [1]
7. 121 [1]
8. $6 \times £1.50 + 4 \times £1.70 = £15.80$ [1]

   $4 \times £15.80$ [1]

   $= £63.20$ [1]
9. 20 [2]

   [1 mark for four cards seen in order with 14 and 18 in
   the middle]

   > The two middle numbers need to be 14 and 18 to
   > give median of 16. 12 + a range of 8 gives 20.

10. 26.2 [2]

    [1 mark for 26 seen]
11. a) $0.4 + 0.15 = 0.55$ [1]

    b) $1 - 0.55$ [1]

    $= 0.45$ [1]
12. ξ

    P   Q

    13

    19   9

    21

    25

    11   15   17   23   27   29

    Fully correct [2]
    [1 mark for 9 placed correctly; 1 mark for all other
    numbers placed correctly with no 'extras']
13. a) $\frac{21}{24} + \frac{16}{24}$ [1]

    $= \frac{37}{24}$ [1]

    $= 1\frac{13}{24}$ [1]

    b) $\frac{4}{5} \times 180$ [1]

    $= \frac{720}{5}$

    $= 144$ [1]
14. a)

x	-1	0	1	2	3
y	$-1\frac{1}{2}$	-1	$-\frac{1}{2}$	0	$\frac{1}{2}$

$-1\frac{1}{2}$ [1]   0   [1]

   b)

   Fully correct [2]
   [1 mark for at least two points plotted correctly]
15. $-\frac{x}{2} > 4$ [1]

    $-x > 8$

    $x < -8$ [1]

    > When you multiply or divide by a negative number,
    > the inequality is reversed.
16. 9.1×10^{-2} [1]

17. a) 10% of 1200 is 120, so 5% of 1200 is 60.

15% = 180 students [1]

1200 + 180 = 1380 students [1]

b) 1380 – 138 (or subtract one-tenth from answer to part (a)) [1]

= 1242 students [1]

18. Shop A: 140p ÷ 5 = 28p [1]

Shop B: 116p ÷ 4 = 29p [1]

Shop A gives the better value for money. [1]

> Use the bus-stop method for division.

19. a) 140 cm ÷ 25 cm [1]

5.6 so 5 complete tiles [1]

> Convert to the same units.

b) 0.6 × 25 **[1]** = 15 cm **[1]**

20. a) 360 ÷ 10 [1]

= 36° [1]

b) 180 – 36 [1]

= 144° [1]

21. a) $\frac{1}{2}$ [1]

b) 0 [1]

22. 200 ÷ 5 **[1]** = £40 per share **[1]**

Ali = £160 (4 × £40), Bob = £240

(6 × £40), Carla = £440 (11 × £40) [1]

> Carla gets five more shares than Bob.

23. 400 grams = 0.4 kg [1]

0.4 × 2.2 [1]

= 0.88 pounds [1]

24. a) Either $\frac{2}{7}$ (Tim), $\frac{11}{31}$ (Ann)

or $\frac{13}{38}$ (combined) [2]

[1 mark for one correct]

b) Either Ann's or combined

depending on answer to part (a). [1]

Reason: More days being recorded

will give more accurate estimate [1]

25. £180 000 = 120% [1]

£180 000 ÷ 120 (× 100) [1]

£150 000 (= 100%) [1]

26. $y = kx$

$3 = k \times 2$

$k = \frac{3}{2}$ [1]

$y = \frac{3}{2}x$

$y = \frac{3}{2} \times 16$ [1]

= 24 [1]

27. $(x + 12)(x + 2)$ [2]

[1 mark for $(x + a)(x + b)$ where $ab = 24$ or $a + b = 14$]

28. $2x + 3y = 3$

$2x - 10y = 16$ [1]

Subtracting equation gives $13y = -13$

$y = -1$ [1]

$x = 3$ [1]

29. Line is $y = 4x + c$ [1]

Substitute $x = -4, y = -9$

$-9 = -16 + c$

$c = 7$ [1]

$y = 4x + 7$ [1]

30. Area of trapezium $= \frac{1}{2} \times (3.5 + 8.5) \times 6$ [1]

= 36 cm² [1]

$xy = 36$ and $x = 4y$

$4y^2 = 36$ [1]

$y^2 = 9$

$y = 3$ [1]

$x = 12$ [1]

Pages 51–64: Practice Exam Paper 2

1. 0.01 [1]

2. $4 - x$ [1]

3. $\frac{1}{6}$ [1]

4. 1.2^3 [1]

5. a)

Empty circle marked at –2 [1]

Directed arrow to the right, from –2 [1]

b) $7x < 28$ [1]

$x < 4$ [1]

6.

Activity	Number of hours
Sleeping	◯◯◯
Eating	◖
Homework	◯
Relaxing	◯◖
Attending school	◯◯

Fully correct [3]

[1 mark for any correct row; 1 mark for 'eating' or 'relaxing' correct]

7. 4 [1]

8. 6 hours 55 minutes [1]

9. $x + (x + 10) + 2(x + 10)$ [1]

$= x + x + 10 + 2x + 20$ [1]

$= 4x + 30$ [1]

10. a) 2, –3, –8 [2]

[1 mark for 2; 1 mark for –3 and –8]

b) Consider $7 - 5n = -57$ [1]

$-5n = -64$

No, since n is not a positive integer [1]

11.

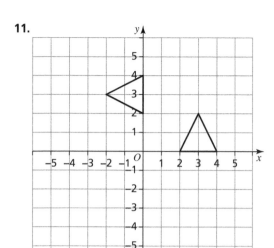

Fully correct [2]

[1 mark for one point of the new triangle correctly plotted]

> When attempting a rotation question, it can help to rotate one vertex at a time, rather than the whole shape.

12. a) Distance = 8.8 cm (accept 8.6–9.0 cm)

8.8×0.25 **[1]** = 2.2 km **[1]**

(accept 2.15–2.25 km)

b) The answer to part (a) would increase. [1]

13. 183.45 cm $\leqslant w <$ 183.55 cm [1]

14. a) $\dfrac{2^2 + \sqrt{4}}{5 + 1} = \dfrac{4 + 2}{5 + 1}$ [1]

$= 1$ [1]

b) Overestimate, numerator values round up and denominator values round down [1]

15. Beans: 14, 28, 42, **56**, 70, 84

Soup: 8, 16, 24, 32, 40, 48, **56**, 64 [1]

Stating a common multiple of 8 and 14, e.g. 56, 112, etc. [1]

4 boxes of beans

7 boxes of soup **[1]** (or any scaled up multiples of 4 and 7)

16. a) $\dfrac{116}{360} \times 100$ [1]

$= 32.2\%$ or 32% [1]

b) The pie chart shows only the proportions of dog breeds, not the actual numbers. [1]

17. a) R J A; R A J; J A R; J R A; A R J; A J R

Fully correct [2]

[1 mark for at least three combinations]

b) 3 identified [1]

$\dfrac{3}{6}$ or $\dfrac{1}{2}$ [1]

18. 7.95 [1]

19. $\dfrac{2}{3}$ [1]

20. $h = \dfrac{P}{dg}$ [1]

$h = \dfrac{20000}{997 \times 9.8}$ [1]

$= 2.05$ m [1]

> Rearrange the equation first, to make h the subject.

21. a) $s = 61°$ [1]

Alternate angles [1]

b) $r = 68°$ [1]

22. Superstrands: £3.59 × 2 = £7.18 [1]

Hair-Care: £3.79 × 3 = £11.37

0.4 × £11.37 = £4.548 [1]

= 11.37 – 4.548 = 6.822 = £6.82 [1]

Hair-Care gives the better value [1]

> Calculate the cost of two bottles at Superstrands, since you get the third for free.

23. 500 g + 1000 g + 1500 g = 3000 g

3000 g ÷ 2170 cm³ **[1]** = 1.38 g/cm³ **[1]**

> Convert all amounts into grams and cm³.

24. a)

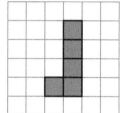

Fully correct [2]

[1 mark for sketching the 2 × 4 rectangle; 1 mark for sketching the 1 × 1 square]

b) 9 cm³ (units must be given) [1]

25. a) £450 × 1.285 **[1]** = €578.25 [1]

b) France: 36 ÷ 1.285 = £28.02 [1]

£28.02 – £27 = £1.02 [1]

26. a)

Time, t seconds	Number of runners	Midpoint	Runners × Midpoint
12 $\leqslant t <$ 13	1	12.5	12.5
13 $\leqslant t <$ 14	5	13.5	67.5
14 $\leqslant t <$ 15	11	14.5	159.5
15 $\leqslant t <$ 16	3	15.5	46.5

[1]

$\dfrac{12.5 + 67.5 + 159.5 + 46.5}{1 + 5 + 11 + 3} = \dfrac{286}{20}$ [1]

$= 14.3$ seconds [1]

b) 4 seconds [1]

> Here the range is given by 16 – 12 = 4

27. $\dfrac{2}{3} \times 162 = 108$ **[1]** go to theme park

$\dfrac{8}{27} \times 162 = 48$ **[1]** go ice-skating

Staying in school are 162 – 108 – 48 **[1]**

= 6 students **[1]**

28. 2 parts = 15 cm

12 parts = 15 × 6 [1]

= 90 cm [1]

29. £56 000 – £12 000 = £44 000

20% of £38 000 = £7600 [1]

40% of £6000 = £2400 [1]

Earnings after tax

= £56 000 – £7600 – £2400 = £46 000 [1]

> Earnings after tax = Salary – Total tax

30. Area large sector $= \frac{3}{4} \times \pi \times 10^2\,(= 235.6)$ [1]

Area small sector $= \frac{1}{4} \times \pi \times 5^2\,(= 19.635)$ [1]

Total area $= \frac{3}{4} \times \pi \times 10^2 + \frac{1}{4} \times \pi \times 5^2$ [1]

$= 255\ \text{cm}^2$ [1]

31. $\sin 63° = \dfrac{x}{18}$ [1]

$x = 18\sin 63°$ [1]

$x = 16.0\ \text{cm}$ [1]

Pages 65–77: Practice Exam Paper 3

1. 150 cm [1]
2. 60 [1]
3. 3500 grams [1]
4. $\dfrac{1}{2}$ [1]
5. Any rhombus drawn, e.g. [2]

[1 mark for a rectangle or a parallelogram drawn]

6. a) 1625.22 kg ÷ 12 = 135.435 kg [1]
135.435 × 15 = 2031.525 kg [1]

> Find the mass of one washing machine first.

b) 2600 kg ÷ 135.435 **[1]** = 19.1974…
19 [1]

> Convert 2.6 tonnes to 2600 kg by multiplying by 1000.

7. 11.5 [1]

> If there are an even number of numbers, then the median occurs exactly halfway between the two middle numbers.

8. a) 48 [1]
b) 3 [1]
c) $y = 3(x + 6)$ [1]
9. 1 : 10 [1]
10. $-1 \leqslant x < 5$ [1]
11. a) $25n + 85t$ [1]
$C = 25n + 85t$ [1]
b) $C = 25 \times 17 + 85 \times 2$ **[1]** = £595 **[1]**
c) $560 = 25n + 85$ [1]

> Subtract 85 from both sides. Then divide both sides by 25.

$475 = 25n$
$n = 19$ [1]

12. $\dfrac{402.5 - 270}{270} \times 100$ [1]

$= 49.1\%$ [1]

13. $a = 3$ and $b = -3$ [1]
14. a) Sum of interior angles
$= 4 \times 180 = 720°$ [1]
$\dfrac{720}{6} = 120°$ [1]
b) The two unmarked angles in the triangle are 60° [1]
So the third angle (y) is
$180 - 2 \times 60 = 60°$ [1]
15. 100 m in 10 seconds = 10 m/s [1]
$10 \times 60 \times 60 = 36\,000$ (m/h) [1]
$36\,000 \div 1000 = 36$ km/h [1]
16. a) 387 000 **[1]** = 3.87×10^5 **[1]**
b) 3.2×10^{-8} [2]
[1 mark for $\dfrac{1}{31250000}$]
17. $35.60 \div 1.28$ **[1]** = 27.8… [1]
$60 - 27.8125 = 32.1875$ [1]
32 [1]
18. a) $9xy(x - 2y)$ [2]
[1 mark for any partial factorisation, e.g.
$9(x^2y - 2xy^2)$]
b) $(x + 4)(x - 4)$ [2]
[1 mark for $(x + a)(x + b)$ where $ab = -16$ or $a + b = 0$]
19. a) 16 **[1]** 26 **[1]**

> This is a Fibonacci-type sequence. Always be on the look-out for these if the sequence does not follow an obvious pattern.

b) $8n - 10$ [2]
[1 mark for $8n + c$]
20. $5 - 2 = 3$, so 3 parts = 9 hours [1]
So 1 part = 3 hours
2×3 [1]
$= 6$ hours [1]
21. a)

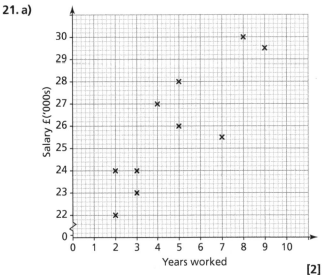

[1 mark for at least three correct plots] [2]

b) Any answer in the range £25 000 – £28 000 [1]
22. Square must come from 16, 25, 36 or 49 [1]
This leaves total of 34, 25, 14 or 1 for
other two so must be total 25 [1]
$25 \div 5 = 5$ and $25 - 5 = 20$

Numbers are 5, 20 and 25 **[1]**

34 ÷ 5 is not a whole number; 14 ÷ 5 is not a whole number; 1 ÷ 5 is not a whole number

23. Factors of 5 or 11 seen **[1]**

$5 \times 5 \times 5 \times 11 \times 11$ **[1]** $= 5^3 \times 11^2$ **[1]**

24. a) $\frac{1}{8} \times 224$ **[1]** $= 28$ **[1]**

b) $\frac{215}{224} \times 100$ **[1]** $= 96\%$ or 96.0% **[1]**

25. a)

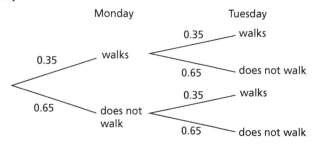

Fully correct **[2]**
[1 mark for any correct branch]

b) 0.65×0.65 **[1]**
   $= 0.4225$ **[1]**

26. $\mathbf{a} - \mathbf{b} = \begin{pmatrix} 4 \\ -3 \end{pmatrix} - \begin{pmatrix} -1 \\ 2 \end{pmatrix} = \begin{pmatrix} 5 \\ -5 \end{pmatrix}$ **[1]**

Yes, correct and e.g. $\begin{pmatrix} -10 \\ 10 \end{pmatrix} = -2 \times \begin{pmatrix} 5 \\ -5 \end{pmatrix}$

Parallel vectors are multiples of each other.

27. a) E **[1]**   b) C **[1]**   c) F **[1]**
28. a) Distance from B to R measured as 9 cm **[1]**
       9×100 m $= 900$ m **[1]**
       [Allow answers in the range 880 to 920 m]
    b) $360° - 45°$ **[1]** $= 315°$ **[1]**   [Allow ± 2°]
    c) Perpendicular bisector of RB drawn. **[1]**
       Correct region shaded **[1]**

Acknowledgements

Every effort has been made to trace copyright holders and obtain their permission for the use of copyright material. The authors and publisher will gladly receive information enabling them to rectify any error or omission in subsequent editions. All facts are correct at time of going to press.

Published by Collins
An imprint of
HarperCollins*Publishers*
1 London Bridge Street
London SE1 9GF

HarperCollins*Publishers*
1st Floor, Watermarque Building,
Ringsend Road,
Dublin 4, Ireland

© HarperCollins*Publishers* Limited 2021

ISBN 9780008326661

First published 2015
This edition published 2021

10 9 8 7 6 5 4 3

All rights reserved. No part of this publication may be reproduced, stored in a retrieval system, or transmitted, in any form or by any means, electronic, mechanical, photocopying, recording or otherwise, without the prior permission of Collins.

British Library Cataloguing in Publication Data.

A CIP record of this book is available from the British Library.

Publishers: Katie Sergeant and Clare Souza
Project Leader: Richard Toms

Authors: Phil Duxbury and Trevor Senior
Videos: Michael White
Cover Design: Sarah Duxbury and Kevin Robbins
Inside Concept Design: Sarah Duxbury and Paul Oates
Text Design and Layout: Jouve India Private Limited
Production: Karen Nulty
Printed and bound in the UK using 100% Renewable Electricity at CPI Group (UK) Ltd

MIX
Paper from responsible sources
FSC www.fsc.org **FSC C007454**

This book is produced from independently certified FSC™ paper to ensure responsible forest management.

For more information visit:
www.harpercollins.co.uk/green